SCIENCE LITERACY

Concepts, Contexts, and Consequences

Committee on Science Literacy and Public Perception of Science

Catherine E. Snow and Kenne A. Dibner, *Editors*

Board on Science Education

Division of Behavioral and Social Sciences and Education

A Report of

The National Academies of
SCIENCES · ENGINEERING · MEDICINE

THE NATIONAL ACADEMIES PRESS
Washington, DC
www.nap.edu

THE NATIONAL ACADEMIES PRESS 500 Fifth Street, NW Washington, DC 20001

This activity was supported by Contract No. HHSN26300077 from the National Institutes of Health. Any opinions, findings, conclusions, or recommendations expressed in this publication do not necessarily reflect the views of any organization or agency that provided support for the project.

International Standard Book Number-13: 978-0-309-44756-0
International Standard Book Number-10: 0-309-44756-9
Digital Object Identifier: 10.17226/23595

Additional copies of this report are available from the National Academies Press, 500 Fifth Street, NW, Keck 360, Washington, DC 20001; (800) 624-6242 or (202) 334-3313; http://www.nap.edu.

Printed in the United States of America

Cover Credit: NASA/ESA, The Hubble Heritage Team (AURA/STScI).

Suggested citation: National Academies of Sciences, Engineering, and Medicine. (2016). *Science Literacy: Concepts, Contexts, and Consequences.* Washington, DC: The National Academies Press. doi:10.17226/23595.

The National Academies of

SCIENCES • ENGINEERING • MEDICINE

The **National Academy of Sciences** was established in 1863 by an Act of Congress, signed by President Lincoln, as a private, nongovernmental institution to advise the nation on issues related to science and technology. Members are elected by their peers for outstanding contributions to research. Dr. Marcia McNutt is president.

The **National Academy of Engineering** was established in 1964 under the charter of the National Academy of Sciences to bring the practices of engineering to advising the nation. Members are elected by their peers for extraordinary contributions to engineering. Dr. C. D. Mote, Jr., is president.

The **National Academy of Medicine** (formerly the Institute of Medicine) was established in 1970 under the charter of the National Academy of Sciences to advise the nation on medical and health issues. Members are elected by their peers for distinguished contributions to medicine and health. Dr. Victor J. Dzau is president.

The three Academies work together as the **National Academies of Sciences, Engineering, and Medicine** to provide independent, objective analysis and advice to the nation and conduct other activities to solve complex problems and inform public policy decisions. The Academies also encourage education and research, recognize outstanding contributions to knowledge, and increase public understanding in matters of science, engineering, and medicine.

Learn more about the National Academies of Sciences, Engineering, and Medicine at **www.national-academies.org**.

The National Academies of

SCIENCES · ENGINEERING · MEDICINE

COMMITTEE ON SCIENCE LITERACY AND PUBLIC PERCEPTION OF SCIENCE

Catherine Snow (*Chair*), Graduate School of Education, Harvard University
Nick Allum, Department of Sociology, University of Essex
John Besley, College of Communication Arts and Sciences, Michigan State University
Dominique Brossard, Department of Life Sciences Communication, University of Wisconsin-Madison
Noah Weeth Feinstein, School of Education, University of Wisconsin-Madison
S. James Gates, Jr., Department of Physics, University of Maryland
Louis Gomez, Graduate School of Education and Information Studies, University of California, Los Angeles
Alexa T. McCray, Division of Clinical Informatics, Harvard Medical School
Janet Ohene-Frempong, J.O. Frempong & Associates, Inc., Elkins Park, PA
Jonathan Osborne, Graduate School of Education, Stanford University
Eugenie C. Scott, National Center for Science Education, Oakland, CA
Earnestine Willis, Department of Pediatrics and Institute of Health and Society, Medical College of Wisconsin

Kenne Dibner, *Study Director*
Julie Anne Schuck, *Associate Program Officer*
Emily Backes, *Research Associate*
Lyla Hernandez, *Director*, Roundtable on Health Literacy
Melissa French, *Program Officer*, Health and Medicine Division
Heidi Schweingruber, *Director*, Board on Science Education
Renée L. Wilson, *Senior Program Assistant*

Preface

In the months leading up to the formation of the Academies' Committee on Science Literacy and Public Perception of Science, the committee's chair had an illustrative conversation with her 4-year-old granddaughter. While walking outside one afternoon, the child reached down and plucked from the ground a small acorn. Looking up at her grandmother, the child brandished the acorn, saying, "Grandma, I'm going to take this home so I can science it."

As our chair's granddaughter is observing here, science is a way of knowing about the world. Just as this child alludes to the notion that the use of science will help her understand something more or better about the acorn in front of her, people have come to depend on science as way to help explain the world around them. In this sense, science can help one understand what the acorn *is*—what it is made of, why it exists—as well as provide the frameworks, standards, and methods within which to evaluate and create new knowledge.

How does science help *everyone,* as citizens of an increasingly interconnected global community, know about the world? Over the course of about 6 months in 2015-2016, the Committee on Science Literacy and Public Perception of Science considered this issue and others in the context of a study on the state of science literacy in the United States. In today's world of immediate access to information and ideas, what is the value of science literacy and how do people know if they have it?

In responding to the charge to the committee, we have written a report that considers how the definition of science literacy has expanded and shifted over time in order to accommodate changing ideas about science. To reflect those changes in the definition of science literacy, the committee has devoted considerable attention to the way that science literacy operates in context. As this report explains, the committee argues that science literacy can be more than just an individual accomplishment and that communities (and societies) can meaningfully demonstrate science literacy in ways that go well beyond the aggregated science literacy of the people in them.

We found this idea particularly compelling at the community level: communities faced with critical environmental issues—such as dangerous levels of lead in their water supply, clusters of cancer or other illnesses, the prospective consequences of fracking, road construction through communities or protected areas, or the introduction of new industries—can achieve levels of sophistication in science literacy that transcend the knowledge or skills of any individual in the community. Under these circumstances, communities not only enable science literacy at the individual level, they also possess science literacy. We devote Chapter 4 to the important phenomenon of community-level science literacy, both to demonstrate that assessing the literacy of individuals does not capture or characterize the emergent literacy of the community and to demonstrate important sources of heterogeneity within nations and societies. Ultimately, it is our goal to move beyond the "horse race" conceptualization of science literacy as a way to judge the state of one nation against others, or to identify individuals who are somehow lacking in knowledge.

Science literacy for individuals, communities, and societies emerges at the interface of the knowledge, attitudes, and motivation of laypeople and the communicative efforts and trustworthiness of scientists. The scientific community needs to take at least partial responsibility for creating an environment in which science literacy can thrive. We envision a society that is infused by science literacy, not in the sense that every person necessarily knows any specific set of things about biology, chemistry, or physics, but in the sense that there is a shared belief that scientific expertise can be trusted, that scientific misconduct and fraud are rare, and that social organizations can and should be structured to enable science literacy rather than prevent it.

Just like our chairperson's granddaughter, we, too, believe that science is one way of knowing about the world. For that reason, we believe it is important that the public understand how scientists work to build increasingly robust explanations by gathering and analyzing empirical evidence. It is our hope that this report offers some insight into how that way of knowing can be an empowering and enabling force for societies, communities, and individuals.

Catherine Snow, *Chair*
Kenne Dibner, *Study Director*
Committee on Science Literacy and Public Perception of Science

Acknowledgments

This report would not have been possible without the contributions of the staff of the National Academies of Sciences, Engineering, and Medicine, the study committee, and many other experts.

First, we acknowledge the sponsorship of the National Institutes of Health (NIH). We particularly thank Carrie Wolinetz from the Office of Science Policy, who provided the committee with insight into how the NIH plans to use this report.

Over the course of the study, committee members benefited from discussion and presentations by the many individuals who participated in our two public meetings. At the first committee meeting, Jon Miller, University of Michigan, provided a thorough explanation of his foundational thinking on defining and measuring science literacy. Dan Kahan, Yale University, provided insight into his research on the relationship between world views (such as political orientation or values) and science literacy. The committee also heard from Philip Kitcher, Columbia University, on the relationship between democracy and science literacy.

At the second meeting, the committee heard from Dietram Scheufele, University of Wisconsin–Madison, on science communication; Ellen Peters, Ohio State University, discussed numeracy; John Durant, director of the museum of the Massachusetts Institute of Technology, and Larry Bell, senior vice president for Strategic Initiatives at the Museum of Science–Boston, talked about the role of informal learning institutions in addressing issues around science literacy.

This report has been reviewed in draft form by individuals chosen for their diverse perspectives and technical expertise, in accordance with procedures approved by the Report Review Committee of the Academies. The purpose of this independent review is to provide candid and critical comments that will assist the institution in making its published report as sound as possible and to ensure that the report meets institutional standards for objectivity, evidence, and responsiveness to the charge. The review comments and draft manuscript remain confidential to protect the integrity of the process. We thank the following individuals for their review of this report: Suzanne R. Bakken, Center for Evidence-based Practice in the Underserved, Reducing Health Disparities Through Informatics, Columbia University; Martin W. Bauer, Department of Social Psychology and Research Methodology, London School of Economics; Rick E. Borchelt, Communications and Public Affairs, Office of Science, U.S. Department of Energy; Anthony Dudo, Stan Richards School of Advertising & Public Relations, Moody College of Communication, University of Texas at Austin; Gerald Gabrielse, Department of Physics, Harvard University; Edith M. Flanigen, consultant, White Plains, New York; Cary Funk, Research on Science and Society, Pew Research Center; Anita K. Jones, School of Engineering and Applied Science, University of Virginia; Ellen Lettvin, Informal STEM Learning, U.S. Department of Education; Bruce V. Lewenstein, Department of Communications, Cornell University; Robert Silbergleit, Department of Emergency Medicine, University of Michigan; and Leon Walls, Elementary Science Education, College of Education and Social Services, University of Vermont.

Although the reviewers listed above provided many constructive comments and suggestions, they were not asked to endorse the content of the report nor did they see the final draft of the report before its release. The review of this report was overseen by Paul R. Sackett, Department of Psychology at the University of Minnesota, and Julia M. Phillips, retired vice president and chief technology officer at Sandia National Laboratories. Appointed by the Academies, they were responsible for making certain that an independent examination of this report was carried out in accordance with institutional procedures and that all review comments were carefully considered. Responsibility for the final content of this report rests entirely with the committee and the institution.

Finally, thanks are also due to the project staff and staff of the Division of Behavioral and Social Sciences and Education (DBASSE). In her first project with the Academies, Kenne Ann Dibner, Board on Science Education, directed the study and played a key role in the report drafting and review process. Emily Backes and Julie Schuck provided critical assistance in project direction, writing, and editing. Renee Gaines Wilson managed the study's logistical and administrative needs, making sure meetings ran efficiently and smoothly. Kirsten Sampson Snyder of the DBASSE staff expertly guided us through the Academies review process, and Genie Grohman provided invaluable editorial direction. Yvonne Wise of the DBASSE staff oversaw the production of the report. Heidi Schweingruber, director of the Board on Science Education, provided infinite wisdom and oversight in this endeavor. Thanks are also due to Health and Medicine Division staff Lyla Hernandez and Melissa French for their guidance and collegiality in the endeavor. The committee also wishes to express its sincere appreciation to the Academies Research Center staff for their support with the comprehensive literature review conducted for this study.

Catherine Snow, *Chair*
Kenne Dibner, *Study Director*
Committee on Science Literacy and Public Perception of Science

Contents

Summary

Science is a way of knowing about the world. At once a process, a product, and an institution, science enables people to both engage in the construction of new knowledge as well as use information to achieve desired ends. Access to science—whether using knowledge or creating it—necessitates some level of familiarity with the enterprise and practice of science: we refer to this as *science literacy.*

Science literacy is desirable not only for individuals, but also for the health and well-being of communities and society. More than just basic knowledge of science facts, contemporary definitions of science literacy have expanded to include understandings of scientific processes and practices, familiarity with how science and scientists work, a capacity to weigh and evaluate the products of science, and an ability to engage in civic decisions about the value of science. Although science literacy has traditionally been seen as the responsibility of individuals, individuals are nested within communities that are nested within societies—and, as a result, individual science literacy is limited or enhanced by the circumstances of that nesting.

In response to a request from the National Institutes of Health (NIH), the National Academies of Sciences, Engineering, and Medicine established an ad hoc committee to study the role of science literacy in public support of science. The study committee, composed of 12 experts across an array of research areas, was tasked with considering existing data about science literacy and health literacy and research on the association of science literacy with public support of science, health literacy, and behaviors related to health. The committee was asked to synthesize the available research literature on science literacy, make recommendations on the need to improve the understanding of science and scientific research in the United States, and consider the relationship between science literacy and support for and use of science and research. In addition, the statement of task guiding this study asked the following questions:

- What is the consensus on metrics for science literacy in the United States?
- What is the evidence on how those measures have changed over time?
- How does this compare to other nations?
- What is the evidence of enhanced science literacy on:
  - Support for, attitudes on, and perception of scientific research?
  - Use of scientific knowledge?
  - Perception of U.S. international standing in science?
  - Health literacy?
  - Behaviors related to health?
- Is lack of science literacy associated with decreased support for science and/or research?

DEFINING AND MEASURING SCIENCE AND HEALTH LITERACY

Science literacy is often construed as knowing the basic facts established by science, but the concept entails much more. We identified three aspects of science literacy common to most applications of the term: content knowledge, understanding of scientific practices, and understanding of science as a social process. We also identified four additional aspects of science literacy that, while less common, provide some insight into how the term has been used: foundational literacy, epistemic knowledge, identifying and judging scientific expertise, and dispositions and habits of mind. Given this range of aspects, it is not surprising that there is no clear consensus about which aspects of science literacy are most salient or important. Different aspects may be more or less important depending on the context.

CONCLUSION 1 The committee identified many aspects of science literacy, each of which operates differently in different contexts. These aspects include (but may not be limited to): (1) the understanding of scientific practices (e.g., formulation and testing of hypotheses, probability/risk, causation versus correlation); (2) content knowledge (e.g., knowledge of basic facts, concepts, and vocabulary); and (3) understanding of science as a social process (e.g., the criteria for the assignment of expertise, the role of peer review, the accumulation of accepted findings, the existence of venues for discussion and critique, and the nature of funding and conflicts of interest).

Though science literacy has been defined in many ways, the aspects highlighted above are some of the most common ideas emerging in the literature, and they represent what some scholars expect would be useful or valuable for individuals using science in their lives, interacting with science information, and making decisions related to science. When considering why science literacy itself would be valuable, some scholars emphasize a personal rationale, defining the term in the context of how science knowledge and knowledge of science can be beneficial to people in their daily lives. Indicators developed to measure science literacy have focused on creating a marker for science knowledge and differentiating between individuals' capabilities.

CONCLUSION 2 Historically, the predominant conception of science literacy has focused on individual competence.

We identify foundational literacy as one aspect of the definition of science literacy. For the purposes of this report, the committee includes numeracy as part of foundational literacy. As such, foundational literacy encompasses the skills and capacities necessary to process and be fluent in the use of words, language, numbers, and mathematics. Domain literacies, like science literacy and health literacy, emerge when a particular set of knowledge or competencies become socially important. The committee recognizes that all domain literacies depend on foundational literacy but may also encompass other skills and knowledge.

CONCLUSION 3 Foundational literacy (the ability to process information—oral and written, verbal and graphic—in ways that enable one to construct meaning) is a necessary but not sufficient condition for the development of science literacy.

Formal definitions of health literacy have developed independently of definitions of science literacy. Because the health literacy field has focused on health behaviors and outcomes, research has examined how health literacy operates in a wide variety of settings and media and has uncovered structural impediments in the health care system. Features of these new, more comprehensive definitions of health literacy, which include aspects such as (1) system demands and complexities as well as individual skills and abilities; (2) measurable inputs, processes, and outcomes; (3) potential for an analysis of change; and (4) linkages between informed decisions and action.

CONCLUSION 4 Concerns about the relationship of health literacy to health outcomes have led to a reconceptualization of health literacy as a property not just of the individual but also of the system, with attention to how the literacy demands placed on individuals by that system might be mitigated.

This reconceptualization of health literacy informed the committee's understanding of science literacy. As a result, the committee supports expanding contemporary perspectives on science literacy to encompass the ways that broader social structures can shape an individual's science literacy. In addition, the committee questions the common understanding that science literacy is, or should be seen only as a property of individuals—something that only individual people develop, possess, and use. Research on individual-level science literacy provides invaluable insight, but it likely offers an incomplete account of the nature, development, distribution, and impacts of science literacy within and across societies. The committee asserts that societies and communities can possess science literacy in ways that may transcend the aggregation of individuals' knowledge and accomplishments. The committee's stance here is relatively new to the field of science literacy: it emerged as a direct result of the opportunity to examine science literacy in relationship to health literacy.

In light of this understanding, the committee organized its thinking about the questions posed in the charge by examining evidence at three levels of science literacy: the society, the community, and the individual. We chose this organization to contrast purposefully with the default understanding of literacy as an individual accomplishment. As a result, the committee chose to delve first into what science literacy looks like at its largest level of social organization—the society.

SCIENCE LITERACY AT THE LEVEL OF SOCIETY

There are four primary rationales for the importance of science literacy: personal, economic, democratic, and cultural. Each of them makes claims about the value of science literacy for nations and societies. Perhaps the most commonly heard claim is that a more science-literate population helps democratic societies make prudent and equitable decisions about policy issues that involve science. Currently, the available evidence does not provide enough information to draw conclusions on whether such claims are justified or not.

Research on science literacy at the level of a nation or whole society can be split into two perspectives. We refer to the first one as the aggregate perspective—empirical work that aggregates data about individuals, usually collected through large public opinion surveys or tests with samples representative of a population, and examines patterns in the whole or by groups. The vast majority of scholarly inquiry at the society level in the field of science literacy, as well as the public discourse, has focused on the aggregate perspective. We refer to the second one as the structural perspective—an alternative way to consider science literacy at the society level by examining the role of social structures. Social structures could include (but would not be limited to) formal policies and institutions (e.g., schools and the scientific establishment) and emergent cultural properties, such as norms of political participation, social and economic stratification, and the presence of diverse groups and worldviews. There is very little research on science literacy from a structural perspective.

Currently, what is measured on science literacy at thc socicty level comes from large public opinion surveys among adults and survey tests of adolescents in many countries. Indicators of adults' knowledge of science are limited to a narrow range of measures on public surveys. It is difficult to draw strong conclusions on cross-national performance from these measures. However, survey responses over time have shown much stability in terms of average performance on knowledge questions, and no country for which there are data consistently outperforms other countries on all questions.

> **CONCLUSION 5** The population of adults in the United States performs comparably to adults in other economically developed countries on most current measures of science knowledge.

The large public opinion surveys in different countries also include measures of attitudes toward science. On these measures, there are many similarities among countries, and response trends have been stable across multiple survey years, particularly in the United States (for which there are more data). The percentage of respondents reporting positive attitudes has been (and remains) quite high, notably in regard to the perceived benefits created by science for societies and support for scientific research.

> **CONCLUSION 6** Current evidence, though limited, shows that populations around the world have positive attitudes toward science and support public funding for scientific research. These attitudes have been generally stable over time. In addition, the same evidence reveals an overall high level of trust in scientists and in scientific institutions.

In reviewing the literature and data from surveys on science literacy, as well as those on foundational literacy and health literacy, the committee found significant disparities in knowledge and access to knowledge. Much more is known about disparities in foundational literacy and health literacy than disparities in science literacy. The committee encourages new research in this area to examine the extent of disparities in science literacy and the social structures that contribute to them.

> **CONCLUSION 7** Within societies, evidence shows that severe disparities in both foundational literacy and health literacy exist and are associated with structural features such as distribution of income and access to high-

quality schooling. Though direct evidence for such structural disparities in science literacy is scarce, we conclude they too exist, in part because the possession of foundational literacy is so integral to the development of science literacy

SCIENCE LITERACY AT THE COMMUNITY LEVEL

Evidence from case studies suggests that science literacy can be expressed in a collective manner—i.e., resources are distributed and organized in such a way that the varying abilities of community members work in concert to contribute to their overall well-being. Science literacy in a community does not require that each individual attain a particular threshold of knowledge, skills, and abilities; rather, it is a matter of that community having sufficient shared capability necessary to address a science-related issue. Examples of such collective capability and action abound.

However, research does not yet show the extent to which communities are able to mobilize to respond to issues at a local level or what features of particular communities enable them to develop and deploy science literacy in effective ways. Evidence from case studies suggests that the success of communities is constrained by structural conditions and depends, at least in part, on the development of scientific knowledge throughout the community and the organization and composition of the community, including the strength and diversity of relationships with scientists and health professionals, scientific institutions, and health systems. The data show that particularly under-resourced communities are more susceptible to the types of environmental and health crises in which science literacy-informed community activism would be crucial, yet they often have the least access to resources that support development and use of science literacy. Additional research is needed to understand the various features and contexts that enable or prevent community science literacy and action.

CONCLUSION 8 There is evidence from numerous case studies that communities can develop and use science literacy to achieve their goals. Science literacy can be expressed in a collective manner when the knowledge and skills possessed by particular individuals are leveraged alongside the knowledge and skills of others in a given community.

The committee also finds that communities can and do contribute to new scientific knowledge in diverse and substantive ways, often in collaboration with scientists. Community involvement has helped to bring new questions to light, provide data that would otherwise be unavailable, encourage the integration of qualitative and observational data with experimental data, increase the robustness and public relevance of data collection strategies, garner political and community support for conclusions, produce new instruments and technologies, and build community awareness and knowledge. Though the evidence describing this phenomenon is still case based, the committee finds that the creation of new scientific knowledge is a compelling demonstration of science literacy.

> **CONCLUSION 9** Based on evidence from a limited but expanding number of cases, communities can meaningfully contribute to science knowledge through engagement in community action, often in collaboration with scientists.

SCIENCE LITERACY AT THE INDIVIDUAL LEVEL

Research on science literacy at the individual level has largely assessed individuals' knowledge using content knowledge assessments and measures of understanding of scientific principles administered through large public surveys. These widely used surveys have provided valuable insight into science knowledge, but constraints on length and demands for comparability over time and across nations mean that they may be limited in what they can capture about science literacy. The existing empirical evidence at the individual level on the value of science literacy is drawn largely from two separate research fields: science literacy and health literacy. Studies on the impact of health literacy have largely examined the relationship between knowledge and behaviors related to health. In contrast, most of the literature on science literacy assesses the relationship between science knowledge and attitudes toward, perceptions of, and support for science.

> **CONCLUSION 10** Research examining the application of science literacy and health literacy has focused on different things: studies on the impact of health literacy have looked for impact on health-related behaviors and actions (e.g., compliance with medical advice, shared decision making, etc.), whereas studies on the impact of science literacy have mostly examined its relationship to individual attitudes toward science and support for scientific research.

Attitudes have been measured by assessing the adult population's evaluation of the social impact of science and technology. These attitudes have been further separated into two groups: a set of broad attitudes toward science and technology that reflect an individual's assessment of the scientific research enterprise generally and a more focused set of attitudes toward specific scientific controversies, such as nuclear power, climate change, stem cell research, and genetically modified foods. Findings demonstrate that context matters when looking at the relationship between knowledge and perceptions of and support for science. Though science knowledge plays a role, many other factors influence an individual's support for science and scientific research.

CONCLUSION 11 Available research does not support the claim that increasing science literacy will lead to appreciably greater support for science in general.

Though there appears to be a small, positive relationship between general science knowledge and general attitudes toward science, scholars have shown that this relationship becomes more complicated when assessing science knowledge and attitudes toward specific science issues. Knowledge impacts diverse sub-groups in the population differently depending on a host of factors, including levels of religiosity, political predispositions and worldviews, and scientific deference. These patterns seem to vary depending on the specific scientific issue being explored and the culture in which the data are collected. In fact, there is often an interaction between knowledge and worldviews such that enhanced knowledge has been associated, in cases of controversial issues, with increased polarization, affecting attitudes toward those specific science issues.

CONCLUSION 12 Measures of science literacy in adult populations have focused on a very limited set of content and procedural knowledge questions that have been asked within the constraints of large population surveys. Though available measures are limited in scope, evidence suggests they are reasonable indicators of one aspect of science literacy, science knowledge. Studies using these measures observe a small, positive relationship between science literacy and attitudes toward and support for science in general.

CONCLUSION 12a An individual's general attitude toward science does not always predict that same individual's attitude toward a specific science topic, such as genetic engineering or vaccines.

CONCLUSION 12b Some specific science issues evoke reactions based on worldviews (e.g., ideology, religion, deference to scientific authority) rather than on knowledge of the science alone.

Research examining the relationship between science literacy, health literacy, and behaviors related to health is limited, but the available examples highlight the weak correlation between science literacy, health literacy, and behaviors. Like the relationship between science knowledge and attitudes toward science, the causal pathway between science literacy, health literacy, and behaviors is complex and mediated by a number of personal and external factors.

These weak relationships suggest that efforts to simply promote knowledge and understanding to change behavior or attitudes may have limited results. Efforts should focus on increasing knowledge while also removing impediments to actions and lowering the literacy demands of particular situations.

CONCLUSION 13 The commonly used measures of science and health literacy, along with other measures of scientific knowledge, are only weakly correlated with action and behavior across a variety of contexts.

MOVING FORWARD THROUGH RESEARCH

The committee offers a conceptualization of science literacy at multiple levels of social organization that is relatively new to the field of science literacy. In order to demonstrate the value of this conception, it will be necessary to develop an evidence base that investigates science literacy in all its complexity.

Recommendation: The committee recommends that, in keeping with contemporary thinking, the scientific community, the research community, and other interested stakeholders continue to expand conceptions of science literacy to encompass (a) an understanding of how social structures might support or constrain an individual's science literacy and (b) an understanding that societies and communities can demonstrate science literacy in ways that go beyond aggregating the science literacy of the individuals within them.

Recommendation: The committee recommends that the research community take on a research agenda that pursues new lines of inquiry around expanding conceptions of science literacy.

The committee notes many places where further research would inform thinking about science literacy. In Chapter 6 we outline a series of research questions as a way of thinking about creating new measures and expanding the information available to better understand. Our questions cover four broad topics: (1) the relationship between science knowledge and attitudes toward science; (2) the utility of science literacy; (3) the relationship of science literacy to other literacy skills; and (4) the role of science literacy for citizens as decision makers.

1

Introduction

The work of science is complex: it is a process, a product, and an institution. As a result, engaging in science—whether using knowledge or creating it—necessitates some level of familiarity with the enterprise and practice of science; we refer to this as *science literacy*. Knowledge of basic science facts is but one small part of the constellation of features that can constitute science literacy. In this report, we document what is known about the components of science literacy, the contexts in which it arises and is used, the foundational literacy and numeracy skills that are prerequisite to it, and the ways in which it is applied, supported, and constrained.

Americans have an ongoing and multifaceted relationship to science. At times in the nation's history, the shifting nature of this relationship has been marked by heightened concern about the ability of Americans to understand, participate in, appreciate, and engage with science, with various stakeholders bemoaning what they perceived to be Americans' decreasing science literacy and worrying about the uncertain future of a citizenry they see as disengaged from or ambivalent toward science.

Despite these episodes of handwringing, the available evidence about the science literacy of the American public does not paint a universally dark picture. As contemporary understandings of science literacy have evolved, so too has the research on what Americans know about science and what they are able to do with that knowledge. This evolution has led to asking and answering questions such as: How should science literacy be defined? How can science literacy be measured? How does science literacy connect to behavior? Is there a connection between science literacy and public support for science? These questions form the background for the committee's study.

COMMITTEE CHARGE AND APPROACH

In response to a request from the National Institutes of Health (NIH), the National Academies of Sciences, Engineering, and Medicine convened an expert committee to examine the role of science literacy in attitudes toward and public support for science, and its relationship to health literacy and health-related behaviors. The specific statement of task for the committee is shown in Box 1-1.

The 12-member committee included experts in several relevant disciplines and areas: science literacy, health literacy, education and learning sciences, international comparisons, survey methods and statistics, and psychometrics and attitude measurement. The committee considered existing data about science and health literacy, research on the association of science literacy with public support of science, health literacy, and behaviors related to health.

BOX 1-1
Committee Statement of Task

The committee will consider existing data about science and health literacy, research on the association of science literacy with public support of science, health literacy, and behaviors related to health. The committee will prepare a final report that synthesizes the available research literature on science literacy and makes recommendations on the need to improve the understanding of science and scientific research in the United States, as well as identifying gaps in our understanding of the relationship between science literacy and support for and use of science and research.

The committee will address the following questions:

- What is the consensus on metrics for science literacy in the United States?
- What is the evidence on how those measures have changed over time?
- How does this compare to other nations?
- What is the evidence of enhanced science literacy on:
  - Support for, attitudes on, and perception of scientific research?
  - Use of scientific knowledge?
  - Perception of U.S. international standing in science?
  - Health literacy?
  - Behaviors related to health?
- Is lack of scientific literacy associated with decreased support for science and/or research?

Interpreting and Addressing the Charge

A major challenge in addressing the charge is the relatively limited array of metrics available for measuring science literacy. As this report describes, the measurements available for cross-national comparisons are thoughtfully developed but limited in scope and depth. Because these assessments are administered to nationally representative samples through the use of costly and labor-intensive surveys, they must be succinct—and, as a result, have often focused on the sort of science knowledge items that can be administered quickly. But many scholars now agree that knowledge of science content falls short of fully representing the construct of science literacy as it is now understood.

In addition, although it is straightforward to document differences across nations or across ethnic groups on those content measures, explaining them is more difficult. It is clear that for some analytic purposes more information is needed—information, for example, about the level of knowledge across multiple domains of science and health, as well as knowledge about the processes scientists engage in and how science epistemology differs from other ways of knowing. The limitations of the commonly used metrics constrain the extent to which the committee can answer the specific questions posed in the charge: in the absence of richer and more complete measures of science literacy, we must often limit our conclusions to what is known about knowledge of an array of science facts and a very limited set of science processes. This report addresses these issues throughout.

Because the charge mandates that this report concern itself chiefly with science literacy,[1] it is the primary lens throughout much of this report. We have attempted as much as possible to differentiate between health literacy and science literacy when the specific point requires it, noting throughout the challenge embedded in teasing health literacy and science literacy apart.

The committee considers health literacy as an important domain that is closely related to and somewhat overlapping with science literacy, though the history and recent developments in the scholarly work on health literacy have been quite different than that on science literacy. Because NIH asked the committee to assist the agency with understanding a potential relationship between science literacy and health literacy, the committee sought research that illuminated the connection across fields: we found few studies. This lack of research made it difficult for the committee to develop an empirically driven discussion of how science literacy and health literacy overlap and how they are distinct. In

[1]The committee notes that research oscillates between the terms "science literacy" and "scientific literacy." The committee cites research and evidence throughout this report that employ both terms. The committee prefers the term "science literacy," but uses "scientific" if specifically quoted. The committee declined to perseverate over the meaning of the specific language, given that research uses the words interchangeably to mean similar ideas.

responding to the statement of task, we use examples from the field of health literacy, as applicable, in order to highlight the overlap across the two fields.

This study was conducted on a notably short timeline of less than 1 year. In order to meet this timeline, the committee elected to address a narrow interpretation of the study charge, which reflects the specific language provided by NIH. As a result, the committee was not able to comment on many interesting and evocative topics that are relevant to the topic of science literacy. There is an unending list of both potential predictors of and consequences for science literacy, and the committee could have proceeded in any number of directions in investigating these ideas. In particular, given committee members' expertise, the committee would have been particularly interested in examining the acquisition of science literacy through both formal and informal education. Similarly, the committee focuses on adult populations and trends in adult data throughout the report. Unfortunately, both time and the specific charge to the committee precluded delving into many topics in depth.

In addition, the committee was mindful that a companion Academies' report on the science of science communication was under way during the time of our investigations. Though the work of the science of science communication study in no way influenced the committee's deliberations, we chose to leave issues related specifically to science communication to that committee. Though the committee is interested in how the institutions of science communicate with the public and the consequences of those interactions, that topic, too, was deemed outside our scope given the time available, the other committee's work, and the specific charge from NIH.

The statement of task specifically asks the committee to make "recommendations on the need to improve the understanding of science and scientific research in the United States." The committee grappled with the underlying assumptions embedded here. Throughout this report, the committee aims to challenge traditional understandings of science literacy, and as a result we note many places at which expanding conceptions of science literacy would require further research. That is, in order to fully understand whether or not there *is* a need to improve the understanding of science and scientific research in the United States, it would first be necessary to solidify an evidence base that investigates science literacy in all its complexity. Again, in order to be responsive to both the charge and the study timeline, the committee did not take on the issue of *how* to improve science literacy, even though that issue is both important and relevant.

In addition to the specific language discussed above, the committee notes a number of places in the statement of task that reflect assumptions about both the concept of science literacy and its utility. These assumptions are particularly noteworthy in the request for the committee to consider evidence of "enhanced scientific literacy on" a list of suggested outcomes, presupposing a relationship between science literacy and those outcomes. Throughout the report, the committee attempts to identify and delineate those assumptions where appropriate while responding to the specific charge from NIH.

Reframing Science Literacy

A first task faced by the committee was to decide how to conceptualize science literacy. We reviewed many definitions and approaches to measurement, considered how those definitions and measurements have changed over time, and catalogued the many aspects of science literacy that have emerged (see Chapter 2 and Appendix A). The committee recognizes that individuals are nested within communities that are nested within societies—and, as a result, individual literacy skills are limited or enhanced by these multiple, nested contexts. In keeping with recent literature on this issue, throughout this report the committee reflects on the ways that social structures might inform the development of an individual's science literacy. Research on individual-level science literacy provides invaluable insights, but on its own offers an incomplete account of the nature, development, distribution, and effects of science literacy within and across communities and societies.

The committee emphasizes another important finding emerging in the literature in its use of the term science literacy in this report: a science literate society is more than the aggregation of science literate individuals. A science literate society or community is a social organization, with traits that can transcend the average knowledge or accomplishments of individuals in that society or community.[2] In light of this broad understanding of science literacy, the committee organized its work to answer the questions posed in the charge by examining evidence at three levels of science literacy: the society, the community, and the individual. We chose this organization to contrast purposefully with the default understanding of literacy only as an individual accomplishment.

FOUNDATIONAL LITERACY

The committee emphasizes that science literacy is the application of foundational literacy skills to a particular domain. Thus it is important to first consider what is meant by "literacy" when it comes with no qualifiers or modifiers. Literacy as a term and a concept has great usefulness and seemingly boundless semantic potential, such that it is used to refer to an ever-larger array of ideas, and the central concept has drifted dramatically from its original meaning. The

---

[2]In a society, people have direct and indirect social connections; in a community, individuals are more closely connected due to shared environments and interests. A community that demonstrates science literacy, for example, might proactively coordinate to monitor whether tap water is potable or could organize to advocate for a specific environmental objective. Chapters 3 and 4 offer in-depth discussions of science literacy at the society and community levels, respectively.

origin is *letra*, Latin for letter, and literacy once very simply referred to the capacity to recognize letters and decode letter strings into recognizable words, along with the concomitant capacity to write words and sentences. That circumscribed meaning has long been transcended, and for the purposes of this report, the committee uses the term "foundational literacy" in reference to the set of skills and capacities described below. The committee asserts that these skills and capacities are effectively foundational to all other domains of literacy, including science literacy.

Even within the field of reading, foundational literacy has been defined in different ways in different historical periods, under different educational policies, through various assessment priorities, and for different segments of the population. The "three Rs" notion of reading that prevailed in the first half of the 20th century was relatively limited—reading mostly meant pronouncing words correctly. That limited notion has been reinforced in public education by efforts to promote grade-level reading, based on the theory that instruction that ensures accurate and fluent decoding by the end of 3rd grade will lead to later comprehension and mastery of other reading literacy challenges (learning from text, synthesizing information from multiple sources, analyzing text to infer the writer's point of view, critiquing claims and arguments in text). In recent years, critics of this approach have argued that the emphasis on that goal in instructional and assessment practices risks diverting attention from the robust developments in reading demands that emerge after 3rd grade, which require instructional attention across the age span and across all subject domains. These demands have now been widely recognized within the reading research community (see Goldman and Snow, 2015).

Even the most conservative of foundational literacy researchers now incorporate a range of extra-textual skills into their notions of literacy. Foundational literacy is commonly extended to include processing words and language in oral contexts, using academic vocabulary and language structures, and having the knowledge base required for comprehension of nontechnical texts about such topics as politics, popular culture, history, art, music, and science. In addition, research on foundational literacy, based as it is in the field of education, has traditionally operated in parallel with research on foundational numeracy, rather than emphasizing the connections between literacy and numeracy. However, this committee asserts that numeracy, defined as the ability to understand probabilistic and mathematical concepts (Peters, 2012), is indeed foundational to other domains of literacy, especially science literacy. Because mathematics represents ideas and concepts in ways that language alone cannot, the committee includes numeracy as part of foundational literacy for the purposes of this report.

All other domains of literacy thus depend on foundational literacy. For science literacy, the production or consumption of science knowledge depends on the ability to access text, construct meaning, and evaluate newly encountered information in the specific domain of science. But the application of the term "literacy" to a specific domain does more than just signify that foundational literacy skills are necessary to understanding the domain itself: it also signifies something like "knowledge, skills, and fluency" within that particular domain. New forms of domain literacy emerge when an individual or group attempts to identify some particular knowledge or competencies as socially important. In other words, framing a domain as an important "literacy" (i.e., media literacy, technology literacy, financial literacy)[3] has become a way of arguing for the importance of ensuring that individuals can access and use the ideas in that particular domain. Not all domain literacies have been the subject of concerted scholarly attention, though, and it is here that science literacy and health literacy stand out: science literacy and health literacy have both emerged as important research arenas, with consequences for policy in a number of contexts.

Finally, the committee notes an important point about the relationship between science literacy and many other domain literacies (in this case, health literacy): health literacy is closely related and somewhat overlapping with science literacy. Science content areas, such as biology or chemistry, are necessary for understanding basic health concepts, and as a result, some science literacy is essential for the knowledge, skills, and fluency necessary to be health literate. As noted above, however, there is relatively little empirical work explaining these relationships, thus limiting the committee's ability to deal in detail with this issue.

EQUITY AND SCIENCE LITERACY

As noted above, the value of science literacy and health literacy—their usefulness and importance to people, communities, and society—is an explicit focus of this report. In order to undergird the committee's arguments about how science literacy and health literacy operate differently in different contexts, it is necessary to raise, at the outset, a critical point about the role of science literacy and health literacy in society: they reflect deep structural inequities in the United States.

Individuals with fewer economic resources and less access to high-quality education have fewer opportunities to develop science literacy and health literacy. This lack of access disproportionately affects some demographic groups: second-language speakers of English, Latinos, black Americans, and children growing up in low-income families or attending under-resourced schools may have fewer opportunities to acquire science literacy (see Chapter 3). Moreover, this inequitable distribution is of particular concern with regard to health

---

[3]See, for example, http://www.medialit.org/reading-room/aspen-institute-report-national-leadership-conference-media-literacy and http://www.mymoney.gov/researcher/Pages/for-researchers.aspx [July 2016].

(Institute of Medicine, 2004). There is strong evidence that health literacy is associated with access to health resources, so those with less opportunity to develop health literacy may as a consequence also experience poorer health care and poorer health outcomes than people with more opportunity to develop health literacy.

At the same time, research from the field of health literacy shows that it would be entirely too simplistic to ascribe poor health outcomes among certain groups exclusively to limitations of an individual's health literacy (Institute of Medicine, 2012). For example, living in a food desert impairs the ability of an individual to gain access to healthy food, regardless of how much they know about the importance of vegetables. Individuals with diabetes may fully understand the mechanisms underlying the disease, but if they are unable to afford regular monitoring of their condition, they are more likely to become sick. In these cases and others, these "undesirable" outcomes cannot be attributed to an individual's deficit of health literacy. Social factors may explain much more of the variability in outcomes than individual levels of health literacy or science literacy. As a result, the committee chooses to emphasize how social factors constrain (or promote) how health literacy and science literacy are expressed at each level of society.

STUDY METHODS

The committee held four in-person meetings and one telephone meeting over the course of the study. The first two were largely information-gathering meetings at which we heard from a variety of stakeholders, including Carrie Wolinetz from NIH's Office of Science Policy, as well as several professional academics with relevant expertise. Jon Miller from the University of Michigan, Dan Kahan from Yale Law School, and Philip Kitcher from Columbia University addressed the committee at its first open session, each speaking to different facets of research on science literacy. At the second open session, Dietram Scheufele from the University of Wisconsin–Madison fielded questions on science communication. Ellen Peters from Ohio State University discussed numeracy. John Durant from the MIT museum, and Larry Bell from the Museum of Science–Boston formed a panel on the role of informal learning institutions in addressing issues around science literacy.

Following those information-gathering meetings, the committee conducted its work in closed session to analyze evidence and formulate conclusions and recommendations. The committee reviewed multiple sources of information in order to consider how science literacy and health literacy may be defined and measured, as well as the relationship between science literacy and the outcomes articulated in the charge.

Multiple fields of research informed the committee's work. Notably, literatures from science communication and science education were considered, as these fields have both proceeded, often in parallel, in attempting to codify what is considered science literacy. Literature from the sociology of science also supported this work. In order to address the health-related components of the charge, the committee reviewed research from the field of health literacy. Literature from psychometrics was considered in order to best synthesize the role of attitude measurement in assessing the potential effects of enhancing science literacy.

The committee also commissioned four supplementary papers intended to support the writing of this report.[4] Lauren McCormack, director for the Center of Communication Science at RTI International, provided a paper on the ways in which health literacy is assessed and measured. Michael Cacciatore, assistant professor of public relations at the University of Georgia, reviewed literature on the role of science literacy in public support for and attitudes toward science and science research. Jon Miller, who spoke to the committee at its first open session, provided a paper on traditional measures of science literacy. Arthur Lupia, professor of political science at the University of Michigan, wrote a paper on science literacy and civic engagement. These papers helped supplement the committee's expertise in order to effectively address the study's statement of task.

The committee expects that this report will be important to a number of groups beyond the study's sponsor. We anticipate that the primary audience for this report will be the science literacy research community, along with science communication practitioners. Science educators (both formal and informal) may be particularly concerned with the committee's discussions about how social structures both constrain and enable the development of science literacy, while policy makers interested in public support for science are likely to find the discussion of the relationship between science knowledge and attitudes toward science informative.

ORGANIZATION OF THE REPORT

The report is organized into six chapters, with two appendices. Following this introduction, Chapter 2 details the history of how science literacy and health literacy have been defined and measured, taking care to note the differences in how the fields of science literacy and health literacy have developed.

Chapter 3 considers science literacy at the society level by summarizing the claims that have been made about how increased science literacy affects societies, considering the role of social structures in science literacy. It also examines how issues at the societal level may constrain science literacy at the community

[4]All commissioned papers may be viewed upon request via the National Academies of Sciences, Engineering, and Medicine's public access file. Jon Miller's paper is also available at this report's National Academies Press website.

and individual levels, and it addresses international comparisons on measures of science literacy.

Chapter 4 examines how communities develop and use health literacy and science literacy and how enhanced literacy in communities may be mobilized to achieve local goals.

Chapter 5 looks at science literacy and health literacy at the individual level, considering how enhanced science and health literacy might affect people: Does it make people more supportive of science? Does it make them better able to use scientific information?

The final chapter offers the committee's recommendations for the field and identifies areas in which new measures and new research inquiries might improve what is known about science literacy and its relationship to support for and use of science and research.

Appendix A presents a table of key definitions and statements about literacy, numeracy, science literacy, health literacy, and health numeracy. Appendix B contains biographical sketches of committee members and staff.

2

Science Literacy and Health Literacy: Rationales, Definitions, and Measurement

The body of research and writing on science literacy is immense and scattered across several scholarly fields. It has also been the subject of several comprehensive reviews (e.g., Miller, 1983, 2004; DeBoer, 2000; Laugksch, 2000; Roberts, 2007; Pardo and Calvo, 2004). In this chapter, we explore the many definitions of both science literacy[1] and health literacy, as well as how the concepts have been measured. In order to put definitions of science literacy in context, we begin by examining some of the common justifications for promoting science literacy because definitions of the term are informed by ideas and assumptions about its value. We then describe how definitions of science literacy have changed over time. Building on this foundation, we then identify a set of aspects that appear to be common across many different definitions in order to provide some clarity about how the term may be both used and understood. We conclude by describing the history of the measurement of science literacy—an enterprise that has remained fairly removed from the conceptual evolution of the term—explaining how the pervasive reliance on narrow measurements of science knowledge constrains understanding of science literacy.

We also discuss the definitions of health literacy, as well as how the concept has been measured. Because science literacy is the primary focus of the committee's charge, we devote most of our attention to this topic, addressing health literacy in separate sections intended to show how the two ideas are—and are

[1]In one effort, Layton, Jenkins, and Donnelly (1994) found 270 meanings and rationales for science and technological literacy.

not—connected. Overall, in this chapter we seek to provide the historical and conceptual context necessary to understand the key arguments in the field and put the following chapters in context.

RATIONALES FOR THE IMPORTANCE OF SCIENCE LITERACY

Four broad rationales have been proposed as to why science literacy is important and necessary: the economic rationale, the personal rationale, the democratic rationale, and the cultural rationale. In this section, we examine each of these rationales in order to provide a context for how the desired outcomes of science literacy inform understanding of the term itself. In addition, we consider the need for science literacy in new media environments (see Box 2-1).

BOX 2-1
Science Literacy in New Media Environments

Though the role of science literacy in relationship to developing technologies and expanding access to information in society is not a separate rationale for science literacy per se, it is indeed an additional, critical concern for why science literacy is both important and necessary in the context of each of the four rationales highlighted above.

Miller (2010) describes an important shift in science learning. While science learners were once expected to "warehouse" science information, increased access to information has changed the predominant model of science learning to the more recent "just-in-time" model. Although media have traditionally been the main source of science information for lay audiences (Nisbet et al., 2002; Dudo et al., 2010), science related questions can now be answered immediately through online searches (Brossard and Scheufele, 2013), signifying less of a need to store content information over time in one's mind (Miller, 2010, p. 192). With increased public access to new science and health information in new media environments, there are opportunities for public access to peer-reviewed studies and other trustworthy science information, but also to information about scientists disagreeing with one another, making mistakes, engaging in fraud, and presuming expertise outside their areas of competence. Before the advent of the Internet, science journalists and mainstream press acted as a kind of natural curb on the dissemination of questionable scientific ideas. In their role as knowledge intermediaries between science and its publics, they acted as gatekeepers and made judgments about the expertise of scientists (Hargreaves and Ferguson, 2000).

Potential access to misinformation has increased as traditional media has de-

The Economic Rationale

The economic rationale for science literacy is closely related to the impetus for educating the general population in science. For instance, a committee set up in the United Kingdom after the World War I to investigate the state of science education argued: "A nation thoroughly trained in scientific method and stirred with enthusiasm for penetrating and understanding the secrets of nature, would no doubt reap a rich material harvest of comfort and prosperity" (Committee to Enquire into the Position of Natural Science in the Educational System of Great Britain, 1918, p. 7). In one form or another, this argument has been a feature of the discussion about the role of science education in society for the past 100 years (see, e.g., Dainton, 1968; European Commission, 2004; Lord Sainsbury of Turville, 2007; National Academy of Sciences, National

clined as the sole source of scientific information. Websites and videos produced by nonscientists and nonscience journalists abound. Though some of these new media productions are reputable sources of science information (Brossard, 2013), others make use of selective evidence to support scientifically questionable views. Some narratives that rival the scientific consensus, such as anti-vaccination arguments, have had tremendous social consequences for public health and individual decision making (Kata, 2010; Poland et al., 2001). Others, such as anti-GMO perspectives, continue to be intensely debated in public dialogue. In short, lay audiences can now make substantial contributions in online environments—accurate or inaccurate—to the body of available information about science.

Moreover, the sharing of personal anecdotes and news stories on social media, as well as the opportunity to "like" stories on Facebook, comments on blogs, and other related features of the current Web 2.0 environment, can also exert a particularly powerful influence on people's attitudes and understanding of science (Brossard, 2013; Goldacre, 2008). Politicians, celebrities, and others who are nonexperts in science, and laypeople with a personal interest in science or health issues, often use social media platforms (such as Facebook or Twitter) for general discussion, sharing information and seeking support from others (Sugawara et al., 2012).

With such a flood of information, science literacy requires the ability to integrate and interpret information, as well as the time and ability for reflection and evaluation (Crowell and Schunn, 2015; Kitcher, 2010; Ryder, 2001). We continue to discuss these needs in relationship to how individuals, communities, and society develop and apply science literacy throughout the rest of this report.

Academy of Engineering, and Institute of Medicine, 2007, 2010; Rutherford and Ahlgren, 1989; The Royal Society, 2014). One of the most recent articulations is offered by Hanushek and Woessman, two prominent economists of education who draw from their extensive analysis of nations' gross domestic product and performance on international tests to argue that the knowledge capital of nations is "powerfully related to long-run growth rates" (Hanushek and Woessman, 2016, p.64).

The essential premise of this utilitarian argument is that advanced economies require a scientifically and technologically skilled population, both in order to fill jobs in science or technology-related professions, such as computer science and engineering, and for the many jobs that require some knowledge of science in today's society, such as nursing, physiotherapy, and construction. Although many authors treat professional training and science literacy as separate goals (e.g., Osborne and Dillon, 2008) proponents of the economic rationale argue that science literacy contributes to professional and economic success across a very wide range of contexts. Science literacy, from this perspective, is a valued outcome because it strengthens economies and economic competitiveness, leading to less unemployment and a high standard of living. Specific to employment claims, however, the mechanisms through which science education contributes to economic growth are contested. For example, countering widespread arguments about the need for more science, technology, engineering, and mathematics (STEM) professionals, data suggest that most STEM fields experience no shortage at all when compared with other professions—computer science and engineering being notable exceptions (Lowell and Salzman, 2007; Salzman, 2013; Weissman, 2013; Xie and Killewald, 2012).

The Personal Rationale

The personal rationale is that science literacy helps people respond to issues and challenges that emerge in their personal and community contexts. According to this rationale, people are confronted with a range of decisions, such as those about health, their consumption of materials and energy, and their lifestyle, in which an understanding of science (or an ability to interact with science) might help them to take informed actions and lead richer, healthier lives (OECD, 2012a). For instance, many conversations with health professionals require some understanding of the body, the structure and function of its many organs and systems, and even the nature of risk. Similarly, decisions and choices about energy may be informed by some understanding of the concept and the consequences of one choice in comparison with another.

The Democratic Rationale

The democratic rationale rests on the claim that a democracy only functions, or at least functions better, when its citizens are informed participants in civic decision making. Proponents of this rationale argue that many of the major problems facing humanity such as the prevention of disease, the production of "clean" energy, the supply of potable water, and climate change—should be understood and addressed at least in part through scientific and technological advances. Only science literate citizens, proponents of this argument claim, are adequately prepared to participate in civic decision making around these challenges. According to a prominent report on education from the European Commission (1995, p. 28):

> Democracy functions by majority decision on major issues which, because of their complexity, require an increasing amount of background knowledge. For example, environmental and ethical issues cannot be the subject of informed debate unless young people possess certain scientific awareness. At the moment, decisions in this area are all too often based on subjective and emotional criteria, the majority lacking the general knowledge to make an informed choice. Clearly this does not mean turning everyone into a scientific expert, but enabling them to fulfill an enlightened role in making choices which affect their environment and to understand in broad terms the social implications of debates between experts.

The democratic rationale revolves around what political and economic theorists call "the commons": goods and resources that are not privately owned. Such goods and resources include the air, oceans, national parks, sanitation, water, public libraries, health infrastructure, and even accumulated scientific knowledge. In a democracy, managing public goods requires active civic engagement to sustain these resources and ensure their equitable distribution and public access. By engaging in such acts as deliberation, persuasion, and the donation of time and money, members of the public participate both in decisions about the use of scientific knowledge (e.g., ways of minimizing air pollution) and decisions about the allocation of resources to the production of scientific knowledge (e.g., supporting funding of stem cell research) (see Rudolph and Horibe, 2015).

The Cultural Rationale

The cultural rationale for science literacy is the idea that the sciences offer some of the "best that is worth knowing" (Spencer, 1884). In the words of Cossons (1993, p. 339):

> The distinguishing feature of modern Western societies is science and technology. Science and technology are the most significant determinants in our culture. In order to decode our culture and enrich our participation—this

includes protest and rejection—an appreciation/understanding of science is desirable.

This rationale is different from those above in that it invokes no extrinsic or utilitarian justification. From this perspective, the sciences are important cultural activities that offer a powerful way of understanding the world and should therefore be part of what it means to be liberally educated (Bereiter, 2002; Committee on the Objectives of a General Education in a Free Society, 1945; Hirsch, 1987; Hirst, 1965; Hirst and Peters, 1970).

Proponents of the cultural rationale point out that science and technology have transformed people's view of the world from one that is flat to one in which it is spherical, where day and night is caused by a spinning Earth instead of a rotating sun, where people look like their parents because every cell carries a chemically coded message about how to reproduce itself, and so on. Although this argument is deeply felt by many scientists and science educators, it is perhaps the least common of the four rationales, and is often obscured by more utilitarian arguments.

TOWARD A DEFINITION

The rationales described above provide context for how the term science literacy is defined. As Norris and colleagues (2014) note, definitions of both science literacy and health literacy invoke a valued direction or desired goal. For instance, the OECD's Programme for International Student Assessment (PISA) report defines science literacy (OECD, 2012a) as the "ability to engage with science-related issues" and undertake "reasoned discourse about science and technology." Such outcomes are not simply an issue of knowing more—rather the outcome is defined by what an individual *might be able to do.* Likewise, Shen's definition of science literacy (which, as we will discuss in following sections, is the rhetorical basis of much of the measurement of science literacy) is not simply knowledge, but rather "the kind of knowledge which can be used to solve practical problems . . . such as health and survival" and a facility that "would bring common sense to bear upon such issues and thus participate more fully in the democratic process of an increasingly technological society" (Shen, 1975b, p. 48). Shen is emphasizing both the personal and democratic rationales for science literacy here, defining the term in the context of how this knowledge will be of benefit. In this section, we explore how shifting ideas about the value of science literacy have informed how the term has been defined.

Definitions of Science Literacy

The term "science literacy" has pervaded much of the public discourse about science education and public understanding since 1958 when it appears to have been coined twice, independently, by Hurd (1958) and McCurdy

(1958), as noted by Laugksh (2000). The phrase was coined as a means of expressing the disposition and knowledge needed to engage with science—both in an individual's personal life and in the context of civic issues raised by both the use of science and technology and the production of more knowledge. Then, as now, there was mounting concern about the growth of science knowledge[2] and the need for the public to engage with the political and moral dilemmas posed by scientific and technological advances. McCurdy (the then president of the Shell Oil Corporation) argued that someone who was science literate would be able to "participate in human and civic affairs." In practice, the term science literacy was used to make an educational case for teaching science to the "90% of all working people" who were not "potential scientists," and who, it was argued, should experience a different kind of science education to enable them to achieve such a goal (Klopfer, 1969, p. 87).

Only 8 years later, the term had become so pervasive that Pella and colleagues (1966) in the Scientific Literacy Center at the University of Wisconsin–Madison identified six distinct types of understanding that were said to be essential to science literacy: the basic concepts in science; the nature of science; the ethics that control scientists in their work; the interrelationships of science and society; the interrelationships of science and the humanities; and the differences between science and technology. Definitions continued to proliferate and become more elaborate over time: 10 years after Pella and colleagues described their six types of understanding, Gabel (1976) constructed a matrix using Pella's categories (now expanded to 8) on one dimension with 9 cognitive and affective objectives on another dimension for a total of 72 separate goals. As Roberts (2007, p. 737) points out: "Thus did scientific literacy become an umbrella concept with a sufficiently broad, composite meaning that it meant both everything, and nothing specific, about science education and the competency it sought to describe."

One of the early definitions that has become influential, at least within the field of measurement of science literacy, is the definition offered by Shen (1975b, p. 46-47), who differentiated three types of science literacy:

- *Practical:* "the kind of knowledge which can be used to solve practical problems . . . such as health and survival."
- *Civic:* "to enable the citizen to become more aware of science and science related issues so that he and his [sic] representatives would bring common sense to bear upon such issues and thus participate more fully in the democratic process of an increasingly technological society."
- *Cultural:* A motivation or "desire to know something about science as a major human achievement."

[2]Rudolph (2002) chronicles the relationship between the launch of Sputnik in 1957 and concern over the state of science education in the United States.

Despite the widespread enthusiasm for science literacy, writ large, and the prominence of a few widely cited definitions, none of the fields concerned with science literacy have managed to coalesce around a common conception of what is meant by the term. Examining the concept 40 years after its inception, DeBoer (2000) identified no fewer than nine overlapping but distinct uses of the term (see Appendix A for more details), and ultimately argued that there was a lack of any universally shared understanding of science literacy other than as "a broad and functional understanding of science for general education purposes and not a preparation for specific or technical careers" (DeBoer, 2000, p. 594).

In the field of education, at least, the lack of consensus surrounding science literacy has not stopped it from occupying a prominent place in policy discourse. From the 1980s onward, science literacy was increasingly presented as a central goal of primary and secondary science education. For example, *Science for All Americans*, a prominent reform document published by the American Association for the Advancement of Science (1989, p. xvii) argued:

> The science-literate person is one who is aware that science, mathematics and technology are interdependent human enterprises with strengths and limitations; understands key concepts and principles of science; is familiar with the natural world and recognizes both its diversity and unity; and uses scientific knowledge and scientific ways of thinking for individual and social purposes.

Ten years later, the UK policy report *Beyond 2000: Science Education for the Future* argued that "the primary and explicit aim of the 5-16 science curriculum should be to provide a course which can enhance 'scientific literacy' " enabling students to, among other things, "express an opinion on important social and ethical issues with which they will increasingly be confronted" (Millar and Osborne, 1998, p. 2009). And although the recently published *A Framework for K-12 Science Education: Practices, Crosscutting Concepts, and Core Ideas* (National Research Council, 2012) avoids the term science literacy, it nevertheless suggests that by grade 12 students should be able to undertake a very similar set of aims—the difference being that these are more specifically defined. Thus, for instance, while students should be able to "read media reports of science or technology in a critical manner so as to identify their strengths and weaknesses" they should also be able to "explain how claims to knowledge are judged by the scientific community today and articulate the merits and limitations of peer review and the need for independent replication of critical investigations" (National Research Council, 2012, p. 73).[3]

Many of these outcomes overlap with the definition offered by PISA, which treats science literacy as a competency—that is, "the ability to engage with

[3]See Appendix A for a complete list of the outcomes defined in *A Framework for K-12 Science Education*.

science-related issues, and with the ideas of science, as a reflective citizen"—in that sense, a concept defined very much by the outcomes of being scientifically literate (Koeppen et al., 2008). According to the OECD (2013, p. 7), a scientifically literate person "is willing to engage in reasoned discourse about science and technology which requires the competencies to:

1. **Explain phenomena scientifically:** Recognise, offer, and evaluate explanations for a range of natural and technological phenomena.
2. **Evaluate and design scientific enquiry:** Describe and appraise scientific investigations and propose ways of addressing questions scientifically.
3. **Interpret data and evidence scientifically:** Analyse and evaluate data, claims, and arguments in a variety of representations and draw appropriate scientific conclusions.

Interestingly, this definition, unlike many others, specifies that the knowledge required to undertake these acts includes not only content knowledge from the various sciences, but also knowledge about how scientists do their work and knowledge about how to make sense of science. While earlier definitions of science literacy sometimes focused on a simplified vision of scientific epistemology referred to as "the scientific method" (Rudolph, 2005; Windschitl, Thompson, and Braaten, 2008), these more recent documents evoke the iterative and social nature of scientific work, emphasizing practices like argumentation and model building in addition to the formulation and testing of hypotheses. After conducting a systematic search of the literature on science literacy, Norris et al. (2014) identified 74 articles containing a distinct definition of science literacy that they then sorted into three categories based on the goals and values inherent in them:

1. *States of knowing* to be obtained—the nature and form of knowledge required.
2. *Capacities* to be developed—the form of actions and competencies a scientifically literate individual should be capable of undertaking.
3. Personal *traits* to be acquired, such as a positive attitude toward science and technology.

The definitions varied in the degree to which they emphasized each of these three goals, and, overall, there is no common agreement about the nature and definition of science literacy.

For some scholars, the key elements of science literacy have been neither knowledge nor capacities but, rather, a particular set of dispositions and habits of mind. This category is broad, including such sweeping ideals as open-mindedness, as well as more specific inclinations, such as a commitment to evidence (Norris et al., 2014; Siegel, 1988). For example, the OECD PISA definition of science literacy includes an interest in science and technology, environmental awareness, and valuing scientific approaches to inquiry. The third part of the definition was seen as important because "scientific approaches to enquiry have been highly successful at generating new knowledge" (OECD, 2013, p. 37). Some researchers, such as Shamos (1995), have argued that is far more reasonable to expect people to develop an appreciation for scientific inquiry, along with a sense of how and when scientific ways of gathering and analyzing evidence have proved particularly successful, than it is to expect them to master a wide range of scientific facts and principles. According to Shamos, science literacy schemes such as the *Benchmarks for Science Literacy* (American Association for the Advancement of Science, 1993, p. 151) are "doomed to fail, for at no time in the entire history of U.S. public school education has even this much knowledge of science been expected, or realised, of high school graduates."

Valuing scientific approaches to inquiry, however, does not mean that an individual has to be positively disposed toward all aspects of science or even use such methods themselves. As Rogers warned, as early as 1948, one should not assume that mere contact with science will make people think critically. The critical disposition that is the hallmark of most scientists when approaching their science—and is something that is acquired through long years of practice and is a feature which, with one or two notable exceptions (Goldacre, 2006; Lehrer, 2010) is too often absent from the communication of science and school science education (Henderson et al., 2015). As a society, while people are good at communicating what they know, they may be less good at communicating how they know—in particular, the central role of critique in establishing claims to know (Popper, 1963; National Research Council, 2012).

The methodological challenge to including dispositions within science literacy is that previous research has often examined whether science literacy *predicts* certain attitudes or dispositions. From this perspective, including dispositions in a definition of science literacy borders on tautological, as something cannot be both a necessary element of science literacy *and* a possible outcome of having or using science literacy. Despite this conundrum, the committee elects to include dispositions as a possible aspect of science literacy because it arises so frequently in the research literature. We summarize the various aspects of definitions of science literacy in Box 2-2.

Reading across the most prominent and influential definitions of science literacy, the committee identified elements that are common to many, if not all, definitions. The most basic of these ideas is that science literacy has value to the people who possess it, whether it solves civic and personal problems or makes the world a richer and more fascinating place and that it should be understood in light of that value (Norris et al., 2014). Beyond this generally shared value, we identified a set of seven commonly proposed aspects of individual science literacy (summarized in Box 2-2). Some scholars would exclude one or more of these aspects from their own definitions, and almost all people who study science literacy emphasize some of these aspects more than others.

Fourez (1997) argues for a somewhat different definition of science literacy. First, he introduces a technological component, arguing that science literacy is inextricable from technology (or technological) literacy. Second, he frames the goals of science literacy differently than is done in most other definitions, arguing that it comprises three central aims: individual autonomy, communication with others, and managing and resolving issues and challenges posed by science and technology. Fourez agrees that a basic knowledge of science and technology provides a certain degree of autonomy, but he points out that the question of what knowledge might be needed is key and must be conceptualized in light of the lives and needs of nonscientists. The knowledge that people need, he argued, is that which empowers them to communicate with others about their life situations, increasing their potential to act (Fourez, 1997, p. 906):

> . . . their knowledge gives them a *certain autonomy*[4] (the possibility of negotiating decisions without undue dependency with respect to others, while confronted with natural or social pressures; a certain capacity to communicate (finding ways of getting one's message across); and some practical ways of *coping* with specific situations and *negotiating* over outcomes).

Fourez's argument about the inextricability of science and technology literacy is an important one that deserves discussion. As various scholars have observed, the social problems and challenges that are associated with science in the public mind are often tied to particular technologies that science has made possible (Kleinman et al., 2005). Technological issues are likely to raise social, economic, ethical, and cultural challenges. Understanding and responding to these challenges requires knowledge of both science and technology. For instance, the possibility that Apple and other phone manufacturers may decide to implement a form of encryption on phones requires some basic knowledge of what is meant by encryption but it also raises a number of social and ethical issues about whether and when it is legitimate for any one manufacturer to do so.

A more complex example is provided by new bioengineering technologies, such as the very recently developed gene-editing technology CRISPR/Cas9, which has already raised ethical and regulatory concerns: see, for example, the statement about this technology from the National Institutes of Health (NIH).[5] Clearly, there is some face validity to the claim that science literacy must also include technology literacy. Yet the distinction between science literacy and technology literacy is not well defined and "science literacy" is often used as an

---

[4]Norris (1997) also explored an idea of science literacy rooted in autonomy—what he called "intellectual independence."

[5]See https://www.nih.gov/about-nih/who-we-are/nih-director/statements/statement-nih-funding-research-using-gene-editing-technologies-human-embryos [July 2016].

BOX 2-2
Commonly Proposed Aspects of Individual Science Literacy

Scholars have defined science literacy in various ways, but seven aspects emerge in many of these definitions. (See Appendix A for a description of key literacy definitions by author.) Notably, these aspects are not included because of evidence of their practical value to individuals (or to communities and societies), but because they represent a sort of theoretical common ground. They are, in short, what many scholars *expect* would be useful or valuable. This list includes: foundational literacies, content knowledge, understanding of scientific practices, epistemic knowledge, identifying and judging scientific expertise, cultural understanding of science, and dispositions and habits of mind.

Foundational Literacies

Science literacy depends on the concepts, skills, understandings, and values generalizable to interpreting texts. These "foundational literacies" include numeracy, textual literacy, visual literacy, and understanding of graphs and charts among others (see e.g., Norris and Phillips, 2003). It should be distinguished from "disciplinary literacy," which is the knowledge and skill required to understand the specific forms of specialized texts commonly used in a discipline to communicate to others within that discipline. (See Chapter 1 for a more detailed discussion of foundational literacy.)

Content Knowledge

Science literacy also depends on scientific content knowledge. Although there is disagreement over the scope of knowledge required, and whether it is genuinely important for people to possess a particular canonical set of knowledge (see, e.g., Feinstein, 2011), it is widely claimed that science literacy involves understanding a set of scientific terms, concepts, and facts, which includes new and older scientific developments, textbook principles covered in K-12 education, and topics discussed in the news, among others (Pella et al., 1966; Frank, 1989; Bauer et al., 1994; Ryder, 2001; Norris and Phillips, 2003; Funk and Rainie, 2015).

Understanding of Scientific Practices

Understanding of scientific practices—how scientists do science—is another commonly mentioned aspect of science literacy. Such understanding, such as the ability to design and evaluate scientific inquiry, is thought to be useful for nonscientists in appraising scientific findings and understanding whether a question has been approached scientifically. In considering the importance of understanding scientific practices for science literacy, some scholars have emphasized the knowledge and skills necessary to do science, such as collecting and analyzing data (Pella et al., 1966; National Science Board, 2016), while others have argued for a more general grasp of what scientists do and how to interpret scientific findings (Ryder, 2001; OECD, 2013). Such an understanding includes knowledge of the common procedures of science, such as peer review, double blind trials, the use of the control of variables strategy, ways of reducing error, and the role and weight of scientific consensus.

Identifying and Judging Appropriate Scientific Expertise

According to many scholars, science literacy requires people to make judgments about the expertise of scientists on the basis of their standing and prestige in the scientific community; the role and weight of their publications and their success in competition for research grants; and the appropriateness of their training and credentials (Norris, 1995; Ryder, 2001; Feinstein et al., 2013; see also National Research Council, 2012).

Epistemic Knowledge

Epistemic knowledge is an understanding of how the procedures of science support the claims made by science. Although related to the above dimensions, epistemic knowledge represents a broader picture of the assumptions and principles that underlie scientific work. Such knowledge enables people to explain why scientific results can be believed, why uncertainty is an inherent aspect of science, how the evaluative process of peer review sustains objectivity, how to recognize the boundaries of science and scientific knowledge, and the ways in which scientific knowledge is constructed by a community over time (see, e.g., Ryder, 2001). People who are science literate along this dimension are aware that science is a human enterprise with strengths and limitations, and appreciate the ethics that guide scientists in their work (Pella et al., 1966; Frank, 1989).

Cultural Understanding of Science

Although this dimension of science literacy is less widely discussed, scholars have long argued that people who are science literate understand the tremendous epistemic achievements of science, appreciate of the beauty and wonder of science and the contributions of science to society—what Shen and others have described as a cultural understanding of science. A cultural understanding of science acknowledges the interrelationships of science and society and science and the humanities and recognizes science as a major human achievement (National Research Council, 2012; Pella et al., 1966; Shen, 1975b; Durant et al., 1989; DeBoer, 2000).

Dispositions and Habits of Mind

Some scholars have argued for the centrality of particular dispositions and habits of mind in science literacy on the grounds that these more general dispositions and habits shape how people engage with science in a wide range of circumstances and may be necessary preconditions for the use of other sorts of skills and knowledge. Dispositions that have been proposed include inquisitiveness, open-mindedness, a valuing of the scientific approach to inquiry, and a commitment to evidence (Shamos, 1995; Lehrer, 2010; Norris et al., 2014).

umbrella term encapsulating both (see, e.g., National Research Council, 1996). Given time constraints and lacking a mandate to explore the nature of technology literacy, the committee chose to continue this common practice.

Defining Health Literacy

The focus of the majority of the definitions for health literacy has been on the capabilities needed by individuals to access and understand health information so that they can act on it. For example, in 1998, the World Health Organization defined health literacy as "the cognitive and social skills which determine the motivation and ability of individuals to gain access to, understand, and use information in ways which promote and maintain good health" (World Health Organization, 1998, p. 10). Shortly thereafter, the American Medical Association (1999, p. 553) stated: "Patients with adequate health literacy can read, understand, and act on health care information." Five years later, the Institute of Medicine (2004) published a consensus study on health literacy and focused on the capabilities needed for individuals to make appropriate health decisions.

Eight years later, Sørensen and colleagues (2012) conducted a content analysis of 17 health literacy definitions, observing that the components of the definitions appear to cluster around six primary concepts: (1) competence, skills, and abilities; (2) actions; (3) information and resources; (4) objective—what health literacy should enable someone or something to do; (5) context—the setting in which health literacy might be needed; and (6) time—the period within which health literacy is needed or developed. Based on this analysis, the authors propose the following "comprehensive" definition for health literacy (Sørensen et al., 2012, p. 3):

> Health literacy is linked to literacy and entails people's knowledge, motivation and competences to access, understand, appraise, and apply health information in order to make judgments and take decisions in everyday life concerning healthcare, disease prevention and health promotion to maintain or improve quality of life during the life course.

The authors expand on this definition to propose a conceptual model that encompasses both the "antecedents" (e.g., age, education, socioeconomic status, culture, societal systems) and "consequences" (e.g., risks to patient safety, poorer health outcomes, health costs) of health literacy.

Recently, there has been increasing attention to the social and physical context in which individuals engage in health activities. As Rudd et al. (2012, p. 26) argue, a more comprehensive definition of health literacy must "include both the abilities of individuals and the characteristics of professionals and institutions that support or may inhibit individual or community action." Unlike earlier definitions that focus almost exclusively on personal decision making and action, this definition also incorporates a capacity for individuals to engage in health-related civic matters. Koh and Rudd (2015, p. 1226) note that the "arc of health literacy bends toward population health" and point to an approach to the concept that includes consideration of social organizations and systems as well as individual capacity.

A recent perspective written by members of the National Academy of Medicine's Roundtable on Health Literacy argues that the old consensus on health literacy is being challenged in interesting and productive ways and that the field "needs to come to a new consensus on the components of a definition of health literacy" (Pleasant et al., 2016, p. 1). They note that health literacy is multidimensional and that it operates in a wide variety of settings and mediums. According to the authors of this report, components of a new more comprehensive definition should include four elements: (1) system demands and complexities as well as individual skills and abilities; (2) measurable components, processes, and outcomes; (3) potential for an analysis of change; and (4) a clearer and more empirically sound linkage between informed decisions and action. In order to have a better understanding of how to improve health status among populations, investigators must have available to them measurement tools that are based on a sound multidimensional definition of health literacy (Pleasant et al., 2016).

We note that the third element, "potential for an analysis of change," means that the definition itself must be open to the ways in which it will inevitably evolve. Pleasant et al. (2016, p. 4) wrote:

> [T]he field of health literacy has come to realize that health literacy is malleable and can change for each person, health professional, or health system for a wide variety of internal and external reasons. A definition of health literacy must become open to that change. Doing so will support and allow researchers to begin to explore how and why change in health literacy occurs.

Including this component in the definition compels the field to regularly consider the ways in which health care needs change over time.

As we have noted in our discussion of the many rationales for science literacy, definitions of science literacy invoke a desired goal and are therefore framed by which rationale or rationales (i.e. economic, personal, democratic, or cultural) the definer is prioritizing at the time (Norris et al., 2014). In the case of health literacy, however, the desired goal implicit in the definitions for health literacy cited here is the promotion and maintenance of good health—for individuals, communities, and societies (World Health Organization, 1998). Though the definitions for health literacy cited here have immediate implications for personal (and community and social) well-being, the promotion and maintenance of good health is a necessary precursor to participation in economic, democratic, and cultural systems.

Summary

A comparison of the research on science literacy and the research on health literacy reveals some overlap. The capacity for civic engagement, which has long been a concern for scholars of science literacy, is emerging as a potential component of health literacy. In contrast, science literacy has only recently started to focus in concrete ways on empirical links to decisions and action—a characteristic emphasis of research and writing on health literacy. Both fields are paying increasing attention to social systems and the way they constrain and enable literate action. In summary, although the two constructs have evolved separately, there is some evidence that the researchers and practitioners who deal with science and health are struggling with many of the same challenges.

MEASURING SCIENCE LITERACY

In this section we consider the development of these measurements and how the measurements have *not* evolved at the same pace as the definitions. As a result, the field faces a concept that cannot yet be fully assessed.

The dominant approach to conceptualizing and measuring science literacy in population surveys has arisen out of work by Jon D. Miller and Kenneth Prewitt in the United States (see Miller, 1983, 1998, 2004) alongside collaborators in Great Britain (see Durant et al., 1989). Underlying these efforts appears to have been widespread concern among policy makers and the scientific community that nonscientists were becoming skeptical about the benefits of science and that such skepticism might result in cuts to science funding that would harm the scientific progress that many argue underpins both American and European economic development (Bauer et al., 2007). The results of the U.S. portion of this work have formed the core of a chapter of a biennial report called *Science and Engineering Indicators* (hereafter, *Indicators*) that the National Science Board provides to Congress and the Executive Branch. Scholars have also used the raw data collected for *Indicators* (which is made publicly available) for peer-reviewed research (e.g., Gauchat, 2012; Losh, 2010), and other countries have used many of the *Indicators*' questions for their own national surveys (e.g., Bauer et al., 2012a; National Science Board, 2016).

Miller (2004) has written that the current approach to assessing science inquiry in surveys began when he and Kenneth Prewitt rewrote a question used by the National Association of Science Writers in 1957 (Davis, 1958) for a 1980 report to the National Science Foundation (NSF) and analyses that appeared in a later journal article (Miller, 1983).[6] This question, which continues to be used, involves asking survey respondents to say whether they feel they have a clear

[6]For a more detailed review of this history, see Pardo and Calvo (2002).

understanding of what it means to study something scientifically.[7] If the respondent says "yes," that respondent is then asked to describe this understanding in his or her own words. These responses are coded using standard procedures. Two additional questions were also added in later years to further assess an understanding of what has been called the "scientific approach" (Miller, 1983) and, later, the "nature of scientific inquiry" (Miller, 1998). The first was added in the 1988 *Indicators* survey and seeks to assess a basic understanding of probability using a multiple-choice question (National Science Board, 1989).

The 1995 survey for *Indicators* then added a two-part question aimed at assessing knowledge of science inquiry that had been piloted in a 1992 study for NIH (National Science Board, 1996). This new question first asked respondents a close-ended question about the best way to test a drug and followed it with an open-ended question about why they thought their method was best.[8]

In addition to knowledge of scientific inquiry, the NSF with Miller also added a battery of true/false and multiple-choice questions aimed at assessing knowledge of basic science concepts to the *Indicators* survey in 1988 (National Science Board, 1989). These "Oxford Scale" science knowledge questions were developed in collaboration with researchers in the United Kingdom (see Durant et al., 1989; Evans and Durant, 1995). In another project, the questions were used as part of a multicountry European survey (Bauer et al., 1994). The science concept questions focused on stable, established areas of knowledge that the survey developers believed would continue to be relevant over time; 11 of the original questions continue to be used (National Science Board, 2016). Other countries have also adopted many of these items in various venues (see National Science Board, 2016, Table 7-3), including multiple European surveys between 1989 and 2005 (Bauer et al., 2012a). Box 2-3 lists the process knowledge questions and Box 2-4 lists the factual questions currently in use by *Indicators*.

Conceptually, Miller (2004, p. 273) has generally argued that a scientifically literate citizen is someone who has both a "(1) a basic vocabulary of scientific terms and constructs; and (2) a general understanding of the nature of scientific inquiry." He writes that the focus on scientific constructs emerged out of a focus in standardized testing in the late 1960s and 1970s, while the focus on the nature of inquiry emerged from efforts around the same time to operationalize the idea of a scientific attitude as described early in the 20th century by John Dewey (1934) and in research related to high school in Wisconsin (Davis, 1935; Miller, 1983). In his early work Miller (1983) argued that someone who

[7]Recent research has highlighted that such self-reported knowledge levels have different relationships with attitudes than measures of knowledge gained from quiz-like questions (Ladwig et al., 2012).

[8]Prior to 1988, construct knowledge had sometimes been measured for *Indicators* using self-reports in which respondents were asked to indicate their level of understanding about such issues as radiation or by asking respondents to list benefits and risks associated with specific technologies (Miller, 1983).

BOX 2-3
Process Knowledge Questions Included in *Science and Engineering Indicators*

Understanding of Scientific Study (open-ended)

- When you read news stories, you see certain sets of words and terms. We are interested in how many people recognize certain kinds of terms. First, some articles refer to the results of a scientific study. When you read or hear the term scientific study, do you have a clear understanding of what it means, a general sense of what it means, or little understanding of what it means?
  - o [If respondent indicates a "clear understanding" or "general sense" response] In your own words, could you tell me what it means to study something scientifically? (Formulation of theories/test hypothesis, experiments/control group, or rigorous/systematic comparison.)

Understanding of Probability (multiple choice, yes/no)

- A doctor tells a couple that their genetic makeup means that they've got one in four chances of having a child with an inherited illness.
  - o (1) Does this mean that if their first child has the illness, the next three will not have the illness? (No)
  - o (2) Does this mean that each of the couple's children will have the same risk of suffering from the illness? (Yes)

Understanding of Experiment (open-ended)

- Two scientists want to know if a certain drug is effective against high blood pressure. The first scientist wants to give the drug to 1,000 people with high blood pressure and see how many of them experience lower blood pressure levels. The second scientist wants to give the drug to 500 people with high blood pressure and not give the drug to another 500 people with high blood pressure, and see how many in both groups experience lower blood pressure levels. Which is the better way to test this drug?
  - o Why is it better to test the drug this way? (The second way because a control group is used for comparison.)

Understanding of Scientific Inquiry (composite of above)

- Correctly answer the ***two probability questions*** stated above.
- Provide a theory-testing response to the open-ended question about what it means to scientifically study something *or* a correct response to the open-ended question about ***experiments*** (i.e., explain why it is better to test a drug using a control group) stated above.

SOURCE: Data from National Science Board (2016).

BOX 2-4
Factual Knowledge Questions Included in Science and Engineering Indicators

Factual Knowledge: Questions Used for "Trend" Scale

- The center of the Earth is very hot. (True)
- The continents on which we live have been moving their locations for millions of years and will continue to move in the future. (True)
- Does the Earth go around the Sun, or does the Sun go around the Earth? (Earth around Sun)
  - o (Part B) How long does it take for the Earth to go around the Sun? (One year) (only asked if Part A correct)
- All radioactivity is man-made. (False)
- It is the father's gene that decides whether the baby is a boy or a girl. (True)
- Antibiotics kill viruses as well as bacteria. (False)
- Electrons are smaller than atoms. (True)
- Lasers work by focusing sound waves. (False)

Factual Knowledge: Questions Not Used for "Trend" Scale

- Human beings, as we know them today, developed from earlier species of animals. (True)
- The universe began with a huge explosion. (True)

SOURCE: Data from National Science Board (2016).

is science literate should further know "both general information about the impact of science on the individual and society and more concrete policy information on specific technological issues," (Miller, 1983, p. 35), but this additional dimension has not been a focus of *Indicators* or substantial work in other countries (but see Bauer et al., 2000). Also, drawing on Shen (1975b), the focus on constructs and inquiry is meant to apply primarily to "civic" science literacy, which Miller describes as the type of science knowledge that might be needed by a citizen to take part in public life, including following news in media outlets, such as *The New York Times* (Miller, 2004, p. 274).

Measurement Validity of the Standard Measures

Methodologically, it is important to recognize that the specific questions used to measure knowledge about scientific inquiry and concepts have always been understood by their developers to represent measures of underlying "latent" constructs. The implication is that an entirely different subset of questions could have been selected and would provide similar results, and any such subset could serve as a proxy for the underlying construct. To the extent that this is true, it is not necessary to ask every possible question to capture the underlying construct. All scale development in social science research builds from this idea, which is roughly analogous to surveying a sample of the population instead of the entire population (see DeVellis, 2003).

Efforts to create longer measures (e.g., the 110-question measure proposed by Laugksch and Spargo, 1996) or measures that better capture specific concepts—such as a wider range of scientific knowledge deemed as important knowledge for the public, particularly the methods of inquiry or the procedures for validating knowledge claims—miss the point. As Sturgis and Allum (2006, p. 333), argue:

> Confusing the contents of the measurement instrument with the attitude or trait underlying responses to it is a common mistake among critics of quantitative approaches to [public understanding of science]. But, as Philip Converse has remarked, it does not take much imagination to realise that knowledge of minor facts … are diagnostic of more profound differences in the amount of contextual information citizens bring to their judgments.

An important related issue is that the development of measures for survey use—when the number of questions that can be used is limited by the amount of time one can realistically ask someone to spend completing a survey—requires a higher level of abstraction than the type of measures that might be used in formal education settings. Open-ended questions are also problematic because it is both expensive and time intensive to reliably code the responses. As a result, instruments such as the "nature of science" batteries developed by Lederman (2007) for use with students and teachers are not suitable for survey research in public settings.

One important change to the items that Miller popularized occurred in the 2010 version of *Indicators* when the National Science Board decided to reduce the battery of knowledge questions used to track factual knowledge from 11 to 9 items, removing the questions on evolution and the big bang. The decision to remove these items was based on research that suggested they were effectively assessing religiosity, rather than factual science knowledge among the U.S. population (for a discussion, see National Science Board, 2012, p. 7-20). The proportion of U.S. respondents giving a correct response on these questions was much higher when they were given alternate wording that did not require them to personally endorse evolution: respondents were asked to respond "true" or "false" to "*According to the theory of evolution*, human beings, as we know them today, developed from earlier species of animals," rather than "human beings, as we know them today, developed from earlier species of animals" (National Science Board, 2016). The questions continue to be asked, but they are not included in the composite scale. It is important to note that the potential validity associated with the "evolution" and "big bang" items is culturally dependent, since these items are not always problematic when included in factual knowledge scales in other countries. We also note that in line with Miller's conceptualization of science literacy (and as discussed above), *Indicators* also includes measures of attitudes toward science, as well as interest in science (see Chapters 3 and 5 for further discussion).

More recently, Kahan (2015, in press) suggests that the current *Indicators'* items—including both the factual knowledge questions and the process questions—could likely be combined into a single measure but that the available questions are too easy (also see Pardo and Calvo, 2004). He argues that the standard measures do a reasonable job at differentiating those individuals with low levels of science knowledge from those individuals with medium levels of science knowledge, they seem to be less useful in differentiating people with medium knowledge from people with high knowledge. Kahan (in press) has thus proposed a new measure that includes some of the more difficult NSF items and adds one item from a list of questions used by the Pew Research Center (see, e.g., Funk and Rainie, 2015), as well as several questions from a short form of a well-established numeracy scale (see, e.g., Weller et al., 2013).

Other researchers have also put forward general measures meant to tap specific aspects of science knowledge or science literacy. Most of these, however, have yet to receive substantial additional use by researchers other than those who created the measures. Among the most prominent efforts is the work conducted by Funk and Rainie (2015) for the Pew Research Center, who attempt to construct their own measure of science knowledge. No validation work on this effort appears to have been published to date. In addition, arguing that science literate individuals need to be able to understand what is present in the normal civic discourse, Brossard and Shanahan (2006) created a measure that directly assesses knowledge about the types of scientific terms used most often in the news media.

Recent examples of efforts to assess additional dimensions of scientific understanding include work on a scientific reasoning scale that attempts to capture understanding of key concepts associated with how science works (Drummond and Fischhoff, 2015), as well as a measure to assess the degree to which respondents understand the uncertainty of scientific evidence (Retzbach et al., 2015). (See chapters below for our examination of the relationship between survey measures of scientific knowledge and desired outcomes such as attitudes toward science or particular sorts of individual action.)

Issue-Specific Measures

In addition to general science knowledge, it is also easy to imagine an infinite range of measures aimed at capturing knowledge of specific scientific areas or domains. However, whereas the public health community continues to create a broad range of measures aimed at capturing knowledge about specific health conditions (Boston University, 2016), only limited research of this sort has been done on other aspects of health or science. As Sturgis and Allum (2006) discuss, one of the few frequently used set of measures is 10 true/false questions aimed at assessing people's knowledge of genetics and biotechnology. This measure appears to have been initially developed for the Eurobarometer (European Commission, 1997; Gaskell et al, 2003), but the questions have also been used in other countries, including the United States (e.g., McComas et al., 2014; Priest et al., 2003). There have also been less expansive efforts to assess knowledge about nanotechnology using six questions (Lee et al., 2005) or just two questions (National Science Board, 2012). More recently, efforts have gone into trying to assess knowledge about climate change (e.g., Hart et al., 2015; Kahan, 2015) and energy (e.g., Cacciatore et al., 2012a; Funk and Rainie, 2015). Efforts to create more general environmental literacy measures are also in progress (Shephard et al., 2014; Zwickle et al., 2014), though there is not yet a standard measure.

Measuring Health Literacy

The number of survey-based instruments to measure health literacy has proliferated over the past decade. At least 112 instruments have been developed, suggesting that there is no consensus on which measure to use and no "gold standard." The Health Literacy Toolshed[9] includes information on these instruments, which focus on a broad range of health contexts and specific health conditions, and which measure a variety of competencies. These include pronunciation (20 measures); communication, including listening (5 measures) and speaking (2 measures); numeracy (55 measures); information seeking (39); and skills related to the application and function of health information (19).

Health literacy measures are used in a variety of ways. Clinicians may use them to assess a patient's health literacy level prior to or at the beginning of a health care visit. Researchers may seek to improve health literacy directly, measuring it before or after implementing an intervention, or they may focus on examining the impact of an intervention on behavior, using health literacy as an independent or control variable.

A recent review of health literacy measures notes that the existing measures vary in the "dimensions that they measure and the level of psychometric rigor to exhibit various aspects of validity." (Haun et al., 2014, p. 327) The authors' conclusion is that there still is no "single rigorously validated health literacy measure that addresses the full range of dimensions" (p. 327) that characterizes health literacy. As health literacy definitions evolve to include the demands of health care systems, there is a concomitant need to develop measurement tools that measure not only individual capabilities, but also the demands of

[9]See http://healthliteracy.bu.edu [July 2016].

health materials, the communication skills of health care professionals, and the expectations and assumptions of health care environments (Rudd et al., 2012).

According to a systematic review by the Agency for Healthcare Research and Quality (AHRQ) (Berkman et al., 2004), most studies of adult health literacy used three instruments: the REALM (Rapid Estimate of Adult Literacy in Medicine) (Davis et al., 1993), the TOFHLA (Test of Functional Health Literacy in Adults) (Parker et al., 1995), and the WRAT (Wide Range Achievement Test) (Jastak and Wilkinson, 1984). However, rather than measuring health literacy directly, these instruments were actually measuring aspects of foundational literacy.

The first national assessment of adult literacy that included a specific focus on health was the National Assessment of Adult Literacy Survey (NAAL) in 2003. NAAL measured prose, document, and quantitative literacy in relation to tasks related to health, including clinical preventive health issues and navigation of the health system. The European Health Literacy Project administered a health literacy questionnaire in eight European nations in 2011 (Sørensen et al., 2015). The questionnaire examined self-reported difficulties in accessing, understanding, appraising, and applying information in tasks related to health care decisions, disease prevention, and health promotion. Additional items related to health behaviors, health status, health service use, and health promotion.

THE PATH FORWARD: A SYSTEMIC VIEW

Early in the committee process, members repeatedly questioned the common understanding that science literacy is, or should be seen as, just a property of individuals—something that only individual people develop, possess, and use. The consensus that emerged from these discussions was that research on individual-level science literacy provides invaluable insight, but also that such research, on its own, offers an incomplete account of the nature, development, distribution, and impacts of science literacy within and across societies. Instead, the committee agreed that science literacy can usefully be studied at different levels of social organization, that research on science literacy at the level of individuals should be complemented by efforts to examine how science literacy emerges in communities of people working together, and how both the nature and effects of science literacy cannot be extricated from the social systems within which they are embedded.

This committee's consensus perspective should not be interpreted as a repudiation of the research that examines science literacy among individuals—by far the majority of research that explicitly focuses on science literacy—as this report details (see, in particular, Chapter 5). Given the committee's consensus, it is still possible to see individuals as developing, possessing, or using science literacy. But it also becomes possible to see individuals as nested within communities, contributing their knowledge and skills to collective actions that appear to be science literate, even when any given individual has a very limited knowledge of science. In this light, communities not only enable science literacy at the individual level, they can under certain circumstances be thought of as possessing science literacy themselves. Finally, the committee agreed that it is essential to consider how an individual or community's position in society affects when and how it is important for them to be science literate and what forms of science literacy will enable them to achieve their personal and civic goals.

Because the committee sees the smaller units of analysis as being nested within the larger ones (individuals within communities that are within the social structures of their societies), we have taken the unusual step of beginning with the largest level—the bird's eye view afforded by comparisons across societies and across groups and structures in society. From here, we shift our attention to science literacy at the level of communities, and then we turn our attention to individuals. This report structure means that most of the research from the field of science literacy (the research that may be most familiar to our scholarly readers) appears toward the end of the report (in Chapter 5), while the preceding chapters draw in scholarship from fields where science literacy is a less common frame. One of the primary research challenges that lies ahead—a challenge that is articulated but not addressed in this report—is understanding how these different levels of analysis can fruitfully be brought together.

CONCLUSIONS

The committee's review of the history of the definitions of science literacy reveals a shifting landscape in which science knowledge has emerged as only one component of a larger and more nuanced construct. Health literacy, too, has evolved, in ways that suggest new potential for synergy between research on health literacy and science literacy.

> **CONCLUSION 1** The committee identified many aspects of science literacy, each of which operates differently in different contexts. These aspects include (but may not be limited to): (1) the understanding of scientific practices (e.g., formulation and testing of hypotheses, probability/risk, and causation versus correlation); (2) content knowledge (e.g., knowledge of basic facts, concepts, and vocabulary); and (3) understanding of science as a social process (e.g., the criteria for the assignment of expertise, the role of peer review, the accumulation of accepted findings, the existence of venues for discussion and critique, and the nature of funding and conflicts of interest).

CONCLUSION 2 Historically, the predominant conception of science literacy has focused on individual competence.

CONCLUSION 3 Foundational literacy (the ability to process information—oral and written, verbal and graphic—in ways that enable one to construct meaning) is a necessary but not sufficient condition for the development of science literacy.

CONCLUSION 4 Concerns about the relationship of health literacy to health outcomes have led to a reconceptualization of health literacy as a property not just of the individual but also of the system, with attention to how the literacy demands placed on individuals by that system might be mitigated.

Science literacy looks quite different depending on what is being demanded and from whom. As the committee illustrates in this report, science literacy can no longer be narrowly identified with the abilities (or limitations) of an individual. Using broader conceptualization would expand the ability to analyze science literacy at the community and system or societal level.

Given this expanding understanding of science literacy, it is critical to be able to accurately assess claims about the level and the consequences of science literacy for individuals, communities, and society. If science literacy is more than just the extent of an individual's science knowledge, is the best means of enhancing it a focus on individuals? This report addresses this question in context by investigating the tensions between the shifting definitions of science literacy and the way it has been measured over time and by considering how those metrics have been used to understand whether (or not) science literacy matters for individuals, communities, and society.

3

Science Literacy in Society and the World

As discussed in Chapter 2, there are four rationales for the importance of science literacy. Three of these arguments (economic, democratic, and cultural) make claims about the value of science literacy for nations and societies. Perhaps the most commonly heard claim is that a more science literate population may help democratic societies make prudent and equitable decisions about policy issues that involve science (European Commission, 1995; Rudolph and Horibe, 2015). However, as discussed in this chapter, the available evidence on science literacy at the society level does not provide enough information to draw conclusions on whether these claims are justified. In this chapter, we review available evidence on science literacy from national and international analyses, focusing on what is known about science literacy and its correlates at the macro level, and we identify areas where evidence is lacking. In our review, we consider different countries as different societies, characterized generally by different cultures, level of economic development, and form of governance. Chapter 4 will further delineate the differences between a society and a community.

TWO RESEARCH PERSPECTIVES

Research on science literacy at the level of a nation or whole society can be split into two perspectives. The first relies on data about individuals, usually collected through large public opinion surveys with representative samples of a population. Aggregating these data and examining them as a whole or by subgroup offers insight into broad patterns of attributes such as awareness, knowledge, and attitudes. This sort of analysis might examine, for instance, the relationship between scientific knowledge and attitudes toward scientific research, looking for patterns within and across countries. We refer to this as the aggregate perspective. As we discuss below, the vast majority of scholarly inquiry into science literacy at the whole-society level, as well as much of the public discourse on science literacy, has adopted this perspective. The aggregate perspective treats all participants in the national sample as individuals and their contribution to science literacy at the society level are represented by statistical means or variances.

The second perspective, which the committee thinks has great potential value, focuses on social structures that likely contribute to science literacy, a broad category that could include (but is not limited to) formal policies and institutions (e.g., schools and the scientific establishment) as well as emergent cultural properties such as norms of political participation, social and economic stratification, and the presence of diverse groups and worldviews. This sort of analysis might examine, for instance, how open meeting laws and other governance structures shape the participation of citizens in science-related decisions, how large-scale information networks affect access to science information, and/or how particular subgroups of citizens differ in their perspective on scientific issues. We refer to this as the structural perspective. Though there has been relatively little research on science literacy from a structural perspective, we discuss later in this chapter how such an approach might help move the field forward.

The two perspectives should be understood as complementary rather than competing ways of understanding the nature and effects of science literacy at the whole-society level. On one hand, the knowledge and attitudes of individuals—attributes that are the focus of the aggregate perspective—influence society through social interaction, which is mediated by social structures. For example, if there is no platform for citizens to contribute to governance of scientific matters, what they do and do not understand about science may have little effect on policy decisions. On the other hand, social structures inform what people know, think, and feel. For example, membership in a cultural group (Kahan et al., 2011) or connection to a social network (Brossard and Scheufele, 2013; Southwell, 2013) affects what information people have and how they are most likely to interpret that information. Research that examines the intersection of these two perspectives is in its infancy, and more work is required to articulate how the aggregate science literacy of individuals and the social structures that shape their lives combine to affect important social outcomes.

SCIENCE LITERACY ACROSS SOCIETIES

The available research on science literacy at the macro-level focuses on cross-national comparisons. The number of countries administering national surveys to measure the public's attitudes toward science and general knowledge of science has grown since the 1970s (Bauer et al., 2012b). Table 3-1 lists many of those surveys (Bauer and Falade, 2014; National Science Board, 2016). These surveys ask standard questions[1] to nationally and internationally representative adult samples, typically 1,000 interviews or more. There is no publication currently that compiles results from all of these surveys. However, there are studies that make comparisons across some countries on similar measures (e.g., see summaries from National Science Board (2016) discussed later in this chapter).

It is possible to compare different countries' aggregated responses to identical (or substantially similar) questions regarding science knowledge and/or attitudes toward science. That said, the committee recognizes the limitations of these comparisons. In the rest of this section, we summarize the available findings that can be compared from surveys across the globe, drawing primarily from the *Science and Engineering Indicators* compiled by the U.S. National Science Foundation (National Science Board, 2016).

The limitations inherent in international comparisons require careful consideration. The available data includes the use of nearly identical questions across multiple countries in multiple years, which reassures the committee that comparisons of aggregated responses can provide some meaningful insight on public knowledge and attitudes toward science. That said, seemingly identical questions can be affected by language and cultural differences that cause questions to be interpreted differently. The data that are being compared were often collected at different time periods and/or in different contexts (i.e., face-to-face interviews versus paper or online questionnaires). For some countries, there are large gaps between data collection years, and there is much irregularity between the countries as to when data were collected (Table 3-1). Some data may not be made available to researchers. Data for international comparison come primarily from Europe, with only limited recent data from the Asia-Pacific region. Data from Africa and South America are especially rare. In addition, like any survey data, the results are subject to sampling and response errors which may result in the omission of significant portions of the target populations or the inclusion of respondents who were not citizens of the countries in which they were surveyed. Also of note, while science and engineering can be understood as distinct fields, national survey data often do not make this distinction (National Science Board, 2016).

Knowledge of Science

Most surveys of adult populations use a small set of quiz-like questions to measure the public's knowledge of underlying science constructs. The nature

[1]See Table 3-2 for common knowledge questions asked across multiple surveys as well as more detailed information about the questions in Chapter 2.

TABLE 3-1 National Surveys of Public Knowledge and Attitudes Toward Science

Country	Survey Titles and Sponsors
United States	National Science Foundation Public Attitudes Toward and Understanding of Science and Technology, General Social Survey (GSS), GSS Science and Technology Module
Years Administered	1979, 1983, 1985, 1988, 1990, 1992, 1995, 1997, 1999, 2006, 2008, 2010, 2012, 2014
Canada	Ministry of Science and Technology; Public Survey of Science Culture in Canada
Years Administered	1989, 2013
China	China Association for Science and Technology, Chinese National Survey of Public Scientific Literacy; China Research Institute for Science Popularization
Years Administered	1991, 1995, 1997, 2001, 2003, 2007, 2010
European Union	Eurobarometer
Years Administered	1977, 1978, 1989, 1992, 2001, 2005
India	National Council of Applied Economic Research
Years Administered	2004
Japan	National Institute of Science and Technology Policy, Survey of Scientific Literacy
Years Administered	1991, 2001, 2011
Malaysia	Science and Technology Information Center, Survey of the Public's Awareness of Science and Technology
Years Administered	2000, 2014
Russia	Survey of Public Attitudes Toward Science and Technology in Russia
Years Administered	2003
Korea	Survey of Public Attitudes Toward and Understanding of Science and Technology–Korea Foundation for the Advancement of Science and Creativity
Years Administered	2004, 2006, 2008
Argentina	Red Iberoamericana de Indicadores de Ciencia y Tecnología
Years Administered	2003
Brazil	Brazilian National Research Foundation, Research Foundation of the State of Sao Paulo, Brazil
Years Administered	1987, 2003, 2006

TABLE 3-1 National

Country	Survey Titles and Sponsors
Bulgaria	Bulgarian Academy of Science, Institute of Sociology, Sofia
Years Administered	1992, 1996 (also included in Eurobarometer survey 2001, 2005)
France	Centre for the Study of Political Life, SciencePo, Paris
Years Administered	1982, 1988 (also included in Eurobarometer survey)
New Zealand	Ministry of Science and Technology
Years Administered	1997
United Kingdom	Economic and Social Research Council, MORI (British public opinion research company); Office of Science and Technology, London
Years Administered	1986, 1988, 1996, 2000, 2004 (also included in Eurobarometer survey)
15 countries, including the United States	BBVA Foundation International Study on Scientific Culture
Years Administered	2011
11 countries, including the United States	World Values Survey
Years Administered	2010-2014

SOURCE: Data from Bauer and Falade (2014) and National Science Board (2016).

and development of these questions is discussed elsewhere in the report (see Chapter 2). Across the set of knowledge questions (see Table 3-2), scores for individual items vary from country to country, and no country seems to outperform the others on every question (National Science Board, 2016). Table 3-2 shows the proportion of each sample by country that answered specific questions correctly for the most recent survey year available. As shown in this table, the United States' correct response rate is higher than some countries and lower than others. The National Science Board (2016) notes that U.S. performance on science knowledge has been fairly stable across 2 decades. In 2014, a representative sample of adults in the United States correctly answered an average of 65 percent of the knowledge questions used for the trend scale.[2] This score

[2]The questions used for the trend scale across years in the *Science and Engineering Indicators* include the nine factual knowledge questions described in Box 2-4 (see Chapter 2). In 2014, Americans were able to correctly answer an average of 5.8 of the 9 items (National Science Board, 2016).

TABLE 3-2 Correct Response to Factual Knowledge Questions in Physical and Biological Sciences, by Country/Region (in percentage)

	United States (2014)	Canada (2013)	China (2010)	European Union (2005)	India (2004)	Japan (2011)	Malaysia (2014)	Russia (2003)	South Korea (2004)
Survey Information									
N	2130	2004	68,416	26,403	30,255	812 to 984	2653	2207	1000
Error margin	±2.5-3.3%	±2.2%					±2.71%		±3.1%
Physical Science Questions									
The center of the Earth is very hot. (True)	84	93	56	86	57	84	75	NA	87
The continents have been moving their location for millions of years and will continue to move. (True)	82	91	50	87	32	89	62	40	87
Does the Earth go around the Sun, or does the Sun go around the Earth? (Earth around Sun)	76	87	NA	66	70	NA	85	NA	86
All radioactivity is man-made. (False)	72	72	48	59	NA	64	20	35	48
Electrons are smaller than atoms. (True)	51	58	27	46	30	28	35	44	46

Lasers work by focusing sound waves. (False)	50	53	23	47	NA	26	30	24	31
The universe began with a huge explosion. (True)	42	68	NA	NA	34	NA	NA	35	67
Biological Science Questions									
It is the father's gene that decides whether the baby is a boy or a girl.* (True)	59	NA	58	64	38	26	45	22	59
Antibiotics kill viruses as well as bacteria. ** (False)	55	53	28	46	39	28	16	18	30
Human beings, as we know them today, developed from earlier species of animals. (True)	49	74	66	70	56	78	NA	44	64

NOTES: Data are for the most recent year available for each country. European Union data includes Austria, Belgium, Cyprus, Czech Republic, Denmark, Estonia, Finland, France, Germany, Greece, Hungary, Ireland, Italy, Latvia, Lithuania, Luxembourg, Malta, the Netherlands, Poland, Portugal, Slovakia, Slovenia, Spain, Sweden, and the United Kingdom, but does not include Bulgaria and Romania. NA indicates data not available or question not asked.
*China and Europe surveys asked about "mother's gene" instead of "father's gene."
**Japan survey asked about "antibodies" instead of "antibiotics."
SOURCE: National Science Board (2016, Table 7-3).

is nearly identical to the average across surveys since 2001 and a slight increase from the average correct in 1992.

In order to respond to the statement of task, the committee reports the National Science Board's comparison of aggregated responses to identical or similar science knowledge questions asked in different countries. As mentioned above, the committee recognizes the limitations of comparing questions asked in different survey administrations under different contexts, but is compelled to use the best data available.

In 2011, the BBVA Foundation surveyed 10 European countries and the United States on a set of 22 knowledge questions that varied from those traditionally asked.[3] The United States, on average, performed similarly to many of the European countries, and slightly above the European average. In this survey as well as others reported by the National Science Board (2016), science knowledge scores have varied between different European countries, with northern European countries, including Denmark, the Netherlands, and Sweden, tending to score the highest. Notably, in comparisons with performances across European countries, U.S. performance on knowledge questions tends to fall in the middle. In the United States, as well as in Europe, researchers have noted that men, younger adults, and more highly educated people tend to score higher on these questions, and these trends have been fairly consistent across different surveys and years (National Science Board, 2008, 2016; Wellcome Trust, 2013).

Variations in level of science knowledge among countries can potentially be attributed to different levels of education across societies and levels of economic development. Bauer (2009) draws on large-scale surveys of knowledge and attitudes across European and Indian states (over 30,000 interviews in each) to conclude that people's knowledge of science does scale with economic development, as indicated by Gross Domestic Product per capita.

Beyond factual knowledge, reasoning capabilities and understanding of the processes of science have not been the focus of surveys from most other countries in recent years. However, the National Science Board (2016, pp. 7-54, 7-55) acknowledges a few available findings in this regard:

> In Asia, a 2010 Chinese survey reported that 49% understood the idea of probability, 20% understood the need for comparisons in research, and 31% understood the idea of "scientific research" (China Research Institute for Science Popularization, 2010). In a July 2011 Japanese survey, 62% correctly answered a multiple-choice question on experiments related to the use of a control group, whereas 57% answered correctly in a follow-up December 2011 survey (National Institute of Science and Technology Policy, 2012). [In the United States], 66% of Americans provided a correct response to a similar question in 2014.

There are also surveys of adolescents' understanding of science, which have

[3]See http://www.fbbva.es/TLFU/dat/Understandingsciencenotalarga.pdf [June 2016].

been used in global assessments of science knowledge and competencies in school-age children. We present this data here because adolescent data is often cited in popular media as an indication of flagging U.S. science achievement. The committee declines to draw conclusions based on this adolescent data, however, as a thorough analysis of this data is outside the scope of this report. As noted in the introduction to this report, we focus on evidence pertaining to adults in order to remain consistent, as much of the research on public perceptions of science uses indicators of adult science knowledge (see Chapter 5).

The most well-known measure is the Programme of International Student Assessment (PISA) from OECD, which was set up in 1997 with the first survey results published in the year 2000. Every PISA survey tests "reading, mathematical and scientific literacy in terms of general competencies, that is, how well students in economically developed countries can apply the knowledge and skills they have learned at school to real-life challenges."[4] PISA does not test how well a student has mastered a school's specific curriculum. The goal of PISA is the measurement of the science literacy of 15-year-old students. In the initial tests, 28 OECD countries, including the United States, and 4 non-OECD countries participated. In 2018, 34 OECD countries and 46 non-OECD countries will participate. The assessment uses a randomized sample of students drawn from each country. The tests last for two hours with one subject out of reading, mathematics, and science defined as the major subject and the two others defined as minor subjects.[5] The tests contain a set of trend items which enable comparability from year to year. The results for the United States in comparison to the lowest and highest achieving countries are shown in Table 3-3. The U.S. performance in science is not statistically different from the average among a wide range of countries. However, enormous variation within the United States has been observed; for example, students from Massachusetts (for which there was a separate sample in 2009) did as well as the Asian countries (Carnoy and Rothstein, 2013).

The other international test in which the United States participates is the Trends in Mathematics and Science Survey (TIMSS). Established in 1995, this measures 8th graders' mathematics and science knowledge every 4 years. However, unlike the PISA, TIMSS is designed to align with each participating country's science curricula. The results of TIMSS "suggest the degree to which students have learned mathematics and science concepts and skills *likely to have been taught in school.*" Further, TIMSS also "collects background information on students, teachers, schools, curricula, and official education policies to allow cross-national comparison of educational contexts that may be related to

[4]See PISA FAQ at https://www.oecd.org/pisa/aboutpisa/pisafaq.htm [July 2016].

[5]What PISA measures is defined by the assessment framework which is rewritten when it is the turn of one of the three major subjects to be assessed. The one defining what was to be assessed in science in 2015, when science was the major subject, was rewritten in 2012 (OECD, 2012a).

TABLE 3-3 International Performance on PISA

Date	Lowest Achieving Countries	U.S. Score	Highest Achieving Countries
2000	Peru: 333 (4.0) Brazil: 375 (3.3) Albania: 376 (2.9) Indonesia: 393 (3.9)	499 (7.3)	Korea: 552 (2.7) Japan: 550 (5.5) Hong Kong: 541 (3.0) Finland: 538 (2.5)
2006	Kyrgyzstan: 322 (2.9) Qatar: 349 (0.9) Azerbaijan: 382 (2.8) Tunisia: 386 (3.0)	489 (4.2)	Finland: 563 (2.0) Hong Kong: 542 (2.5) Canada: 534 (2.0) Taiwan: 532 (2.6)
2012 (Science was not a major subject this year)	Brazil: 405 (2.1) Kazakhstan: 425 (3.0) Costa Rica: 429 (2.9) Cyprus: 438 (1.2)	497 (3.8)	Shanghai-China: 580 (3.0) Singapore: 551 (1.5) Hong Kong: 555 (2.6) Japan: 547 (3.6)

NOTE: The scores are normalized to a scale which goes from 0-1000 with a mean of 500 and standard deviation of 100. Standard errors appear in parentheses.
SOURCE: Data from OECD and UNESCO Institute for Statistics (2003) and OECD (2007, 2014).

student achievement" (Provasnik et al., 2012, p. 1[emphasis added]). The committee therefore sees these results as a way to measure how well an education system performs an international context, rather than as a measure of science literacy. See Table 3-4 for TIMSS results.

Attitudes toward Science

According to the National Science Board (2016, p. 7-6), "...Americans' overall attitudes about science are either stable or becoming more positive" and are generally comparable to those of other countries. For example, spending on basic scientific research has been consistently supported by U.S. public opinion. About 80 percent of U.S. respondents have agreed "that the federal government should fund scientific research" (p. 7-7) across multiple survey years from 1981 to 2004. Levels of agreement in South Korea, Malaysia, Japan, and Brazil are similar to those in the United States, whereas agreement is slightly less in Canada, China, and Europe at 72-77 percent (National Science Board, 2016).

Several different surveys have asked similar questions about the perceived past and future benefits of science and technology (S&T) (European Commission, 2013; Council of Canadian Academies, 2014; World Values Survey, 2014). Responses across countries have been largely favorable with the public viewing positive influences of science on society and prospects for the future (see Box 3-1 for comparisons of different countries).

In a similar question, the World Values Survey asked respondents to address whether science has made the world better off or worse off (World Values Survey, 2014); most respondents across countries agreed the world was better off because of science (National Science Board, 2016). Among OECD countries surveyed 2010-2014, the U.S. response was similar to those in other countries with 79 percent agreeing the world was better off, compared to responses in South Korea (84%) and Sweden (80%), and Japan (75%).

Data are available for the U.S. public on people's reported confidence in the leaders of the scientific community. Here, again, there has been much stability in views since the 1970s. In 2014, 90 percent of U.S. respondents expressed a great deal or some confidence in leaders in the scientific community, rating science more favorably than almost all other institutions (see Figure 3-1).

TABLE 3-4 International Performance of 8th-grade Students on TIMSS and the Highest- and Lowest-performing Countries

Date	Lowest Achieving Countries	U.S. Score	Highest Achieving Countries
1995	South Africa: 326 (6.6) Columbia: 411 (4.1) Kuwait: 430 (3.7) Cyprus: 463 (1.9)	534 (4.7)	Singapore: 607 (5.5) Czech Republic: 574 (4.3) Japan: 571 (1.6) Korea: 565 (1.9)
1999	South Africa: 243 (7.8) Morocco: 323 (4.3) Phillipines: 345 (7.5) Chile: 420 (7.8)	515 (4.6)	Taiwan: 569 (4.4) Singapore: 568 (8.0) Hungary: 552 (3.7) Japan: 550 (2.2)
2003	South Africa: 244 (6.7) Ghana: 255 (5.9) Botswana: 365 (2.8) Phillipines: 377 (5.8)	527 (3.1)	Singapore: 578 (4.3) Taiwan: 571 (3.5) Korea: 558 (1.6) Hong Kong: 556 (3.0)
2007	Ghana: 303 (5.4) Qatar: 319 (1.7) Botswana: 355 (3.1) El Salvador: 387 (2.9)	520 (2.9)	Singapore: 567 (4.4) Taiwan: 561 (3.7) Japan: 554 (1.9) Korea: 553 (2.0)
2011	Ghana: 306 (5.2) Morocco: 376 (2.2) Indonesia: 406 (4.5) Lebanon: 406 (4.9)	525 (2.6)	Singapore: 590 (4.3) Taiwan: 564 (2.3) Korea: 560 (2.0) Japan: 558 (2.4)

NOTES: Scores are reported on a scale which is normalized to go from 0-1000 with an average of 500 and standard deviation of 100. Standard errors appear in parentheses.
SOURCE: Data from Beaton et al. (1996) and Martin et al. (2000, 2004, 2008, 2012).

BOX 3-1
International Comparisons on Perceived Future Benefits from Science

For the Eurobarometer, Europeans were asked whether they believe S&T [science and technology] would "provide more opportunities for future generations." Three-quarters of Europeans (75%) agreed, and several northern European countries were again among the most favorable, led by the Netherlands (88%), Estonia (87%), Denmark (85%), and Sweden (85%). The least positive attitudes were in Southern and Eastern Europe, including Slovenia (64%), Romania (67%), and Italy (67%).

Among OECD countries in the [2010-14 World Values Survey], the 79 percent of Americans who said they believe S&T will ensure more opportunities for future generations is similar to results from the Netherlands (84%), South Korea (80%), and Australia (74%). The OECD countries that see the most hope from S&T are Estonia (93%) and Poland (86%). Beyond the OECD, the countries in which there appears to be the most hope for S&T include Libya (97%), Qatar (93%), Uzbekistan (93%), and Armenia (91%) (World Values Survey, 2014). A separate 2013 survey indicated that 74 percent of Canadians agreed that S&T would create more opportunities for the next generation (Council of Canadian Academies, 2014).

. . . most [Asian] respondents appeared to support S&T. In 2010, 75% of Chinese respondents "fully" or "basically" agreed that S&T brings more advantages than disadvantages . . ., (China Research Institute for Science Popularization, 2010). In 2011, 54% of Japanese respondents said that S&T development has more advantages than disadvantages (National Institute of Science and Technology Policy, 2012). South Koreans were asked separate questions about the risks and benefits of S&T. In 2012, about 83% "agreed" or "somewhat agreed" that S&T promotes a healthy and convenient life, and 72% agreed that S&T "helps in everyday life." However, 60% also agreed that S&T "creates problems" (Korean Foundation for the Advancement of Science and Creativity, 2013).

SOURCE: National Science Board (2016, pp. 7-61, 7-62, 7-65).

The relationship between science knowledge and attitudes toward science for individuals is discussed in detail in Chapter 5. However, we consider here what is known across countries (societies) about this relationship. The meta-analysis by Allum et al. (2008, p. 51), synthesizing results from public opinion surveys in a range of countries from 1989 to 2005, found "there is very little cross-cultural variation in the correlation between knowledge and attitudes. . . . " The committee inferred from these results that the mechanism by which science knowledge and general attitudes toward science are associated, whatever it may be, is likely very similar in all countries for which we have knowledge and attitude measures. This weak positive correlation is discussed further in Chapter 5.

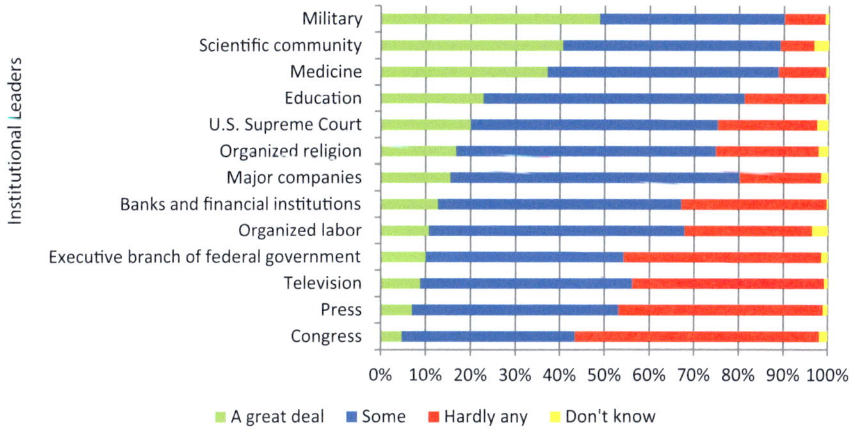

FIGURE 3-1 Public confidence by type of institution, 2014.
NOTE: The responses were to the following question: "As far as the people running these institutions are concerned, would you say that you have a great deal of confidence, only some confidence, or hardly any confidence at all in them?"
SOURCE: National Science Board (2016, Fig. 7-17); data from the 2014 General Social Survey.

Others have looked at cross cultural variations in attitudes/support for specific science innovations; one study looked at nanotechnology in the United States and Singapore (Liang et al., 2013). The authors concluded that the mechanisms at play were different in the two nations when a specific scientific issue is considered. In each country, perceived familiarity with nanotechnology and of the benefits and risks of the emerging technology tend to be interpreted differently. The responses were filtered through different values (religiosity and deference to scientific authority), therefore indirectly affecting public support in different ways (Liang et al. 2013).

Research using the country as the unit of analysis is rare. However, this type of research may be helpful in understanding the impact of variations in science literacy at the society level. One exception is the work of Durant and colleagues, who proposed that societies or countries with higher levels of scientific, technological, and industrial development (as measured by standard international indices) will attend to science knowledge differently than those societies or countries with lower levels (Durant et al., 2000). Using data from the 1992 Eurobarometer with respondents in 12 European countries, the researchers found that the level of public scientific knowledge (represented by the aggregated mean of individual measures) increases as the level of industrialization[6] increases. Furthermore, they found that "sociodemographic variables[7] are better predictors of knowledge in the less industrialized countries than in the advanced industrialized countries" (Durant et al., 2000, p. 141). With the 1992 Eurobarometer data, they also observed that average interest in science increases among countries from low- to mid-levels of industrialization and that average interest drops off for countries with the highest levels of industrialization. The researchers suggest that this trend may occur because more knowledgeable societies may take science knowledge for granted and as a result show lower levels of interest in science. A similar pattern emerged in comparisons of average attitudes toward science in Indian states (with relatively low economic development) and European states (with relatively high economic development) (Bauer, 2009), as well as in comparisons of China to Europe (Liu et al., 2012).

Interpreting Data on International Comparisons

Cross-national research is useful for revealing broad-brush patterns and trends over time. In recent decades, such comparisons have focused on performance metrics. For countries for which there are data, attitudes toward science have been (and remain) quite positive, particularly in regard to support for scientific research and opportunities created by science for societies. Across indicators of science knowledge on public surveys of adults, scores for individual questions vary from country to country, and no country consistently outperforms other countries. There is some evidence that society-wide levels of educational attainment and economic development correlate with average scores on survey measures of science knowledge. There has also been some research examining variations in the relationship between knowledge and attitudes across different societies, but more work is needed to draw strong conclusions.

Some observers interpret these comparisons and scores as reassuring, given the relative similarity across countries. Others see them as disheartening, on the grounds that populations in developed countries should score higher on measures of science knowledge.[8] At times, findings on individual knowledge questions in national surveys have been taken out of the context of gauging

[6]Durant and colleagues use a socioeconomic indicator of industrialization developed by Bairoch (1982).

[7]The term "sociodemographic" is used in the European literature and refers to variables involving social and demographic factors, such as gender, age, level of education, and income class.

[8]See, for example, http://www.pewresearch.org/fact-tank/2015/02/02/u-s-students-improving-slowly-in-math-and-science-but-still-lagging-internationally/ or http://blogs.discovermagazine.com/loom/2012/06/11/science-literacy-a-worldwide-look/#.V1BeyU3ruUk or https://www.sciencedaily.com/releases/2007/02/070218134322.htm [June 2016].

science literacy from a set of science questions and over-interpreted.[9] Perhaps more importantly, the historical focus on *average* levels obscures the very considerable diversity within societies. The next section examines how science literacy (and the related concepts of foundational literacy and health literacy) is distributed in the United States.

VARIATIONS IN LITERACY WITHIN SOCIETY

When it comes to measures of science literacy, there is often greater variation within a country than between countries. In other words, the difference between the most and least knowledgeable people in a society is far greater than the knowledge difference between the average citizen and the average citizen in another country (Sum et al., 2002; Carnoy and Rothstein, 2013). A closer look at variation within one country (the United States) reveals stark disparities in knowledge, access to knowledge, and access to systems that enable people to interpret and act on the knowledge they have.

Research on stratification, variation, and disparities in science literacy is very rare. Existing data on science knowledge, of the sort collected in the ongoing *Science and Engineering Indicators* project, offers limited insight. The subgroup differences in the *Science and Engineering Indicators*[10] are available by age, gender, education level, and family income. Surveys of factual knowledge since the early 1980s tend to show a significant gap between the top-performing age group (ages 25-34 in the 2014 results) and older age groups, though this gap is narrowing. The overall average score for men tends to be higher than for women (69% and 61%, respectively, in 2014); however, differences depend on specific questions asked. In addition, a strong relationship has been observed between scores of factual knowledge and level of formal schooling and the number of science and mathematics courses a person has completed. Information on family income has been compared since 2006. Respondents in the top income quartile have consistently scored higher than those in the bottom quartile (76% and 54%, respectively, in 2014) (National Science Board, 2016).

A recent Pew Research report, based on a survey of knowledge on 12 science-related questions (see Box 3-2),[11] finds similar relationships between demographic factors and science knowledge (Pew Research Center, 2015). Generally, younger adults displayed slightly higher overall knowledge of science

---

[9]See, for example, http://www.science20.com/news_articles/science_literacy_american_adults_flunk_basic_science_says_survey-47608 [July 2016].

[10]Data reported in the *Science and Engineering Indicators* are taken from relevant questions asked on the General Social Survey: see Appendix Table 7-6 in National Science Board (2016) for a breakdown of percentage answering trend factual knowledge questions correctly by respondent characteristics.

[11]For more information, see http://www.pewinternet.org/2015/09/10/what-the-public-knows-and-does-not-know-about-science/ [May 2016].

BOX 3-2
Questions from the Pew Research Center Survey

Below are the questions and the percentages of people answering them correctly.

1. Earth's core is its hottest layer (86%).
2. Uranium is needed to make nuclear energy/weapons (82%).
3. A comet has icy core and tail of gas and dust (photo question) (78%).
4. Ocean tides are created by gravitational pull of moon (76%).
5. Jonas Salk developed polio vaccines (photo question) (74%).
6. Distinguish definition of astrology from astronomy (73%).
7. Radio waves are used to make/receive cellphone calls (72%).
8. A light-year is a measure of distance (72%).
9. Interpret a scatterplot chart (graph) (63%).
10. Identify how light passes through magnifying glass (image question) (46%).
11. Amplitude or height determines loudness in a sound wave (35%).
12. Water boils at lower temperature at high altitudes (34%).

SOURCE: Data from Pew Research Center (2015).

than adults ages 65 and older, though this trend was reversed on some specific questions. The Pew study also found a gap in science knowledge between men and women, with men outperforming women on many questions, even when comparing men and women with similar levels of education. Notably, the questions on the Pew survey were primarily drawn from the physical sciences, an area in which men tend to fare better than women, as contrast to the biological sciences, an area in which women tend to answer more questions correctly (National Science Board, 2016).

The Pew study also found differences associated with race and ethnicity. Whites were more likely than Hispanics or black Americans to answer more of the questions correctly, on average; the mean number of items correct is 8.4 for whites, 7.1 for Hispanics, and 5.9 for black Americans.[12] These findings should be interpreted with caution due to the smaller number of respondents in this survey who are black American (N = 259) or Hispanic (N = 247) compared to the number of respondents who are white (N = 2,551). The findings, though,

[12]The Pew Research Center report (2015, p. 5) points out that "the findings on race and ethnicity are broadly consistent with results on science knowledge questions in the General Social Surveys between 2006 and 2014. Pew Research analysis of the GSS data finds white adults scored an average of 6.1 out of 9 questions correctly, compared with 4.8 for Hispanics and 4.3 for black Americans."

are consistent with those from previous Pew Research surveys[13] as well as other studies that measure knowledge or educational performance across a range of domains.

The Pew Research report is a rare analysis of race and ethnic differences in regards to science literacy. In most studies on scientific knowledge and attitudes data are not disaggregated by race, class, or other social group because of issues such as sample size within the subgroups. However, educational measures assessing students' knowledge, for which sample sizes are often larger, are commonly disaggregated in this way, and they reveal stark differences in scientific knowledge by race and socioeconomic status (SES), as well as geographic differences from state to state and region to region within a state (National Center for Education Statistics, 2012b; U.S. Department of Education, 2014b). Although the results of standardized science assessments in school settings may be an imperfect proxy for adult science literacy, researchers studying knowledge gaps for decades have observed in both cross-sectional and time-series research that people with higher levels of education and science knowledge are significantly more likely to acquire and understand science information from public sources, which suggests that differences in knowledge and interest are likely to expand rather than contract over lifespans (Viswanath and Finnegan, 1996; Southwell, 2013).

In contrast with the scarcity of research on disparities in science literacy, there is a substantial body of research focused on disparities in foundational literacy and health literacy. Taken together the results of this research paint a sobering picture of intra-national disparities. Comparisons of this sort evoke a shallow and problematic deficit perspective[14] that may not accurately portray the competence of particular individuals and groups. At the same time, comparisons of this sort are useful ways of demonstrating inter-group differences on particular measures in ways that prompt discussion about the origins and consequences of those differences.

Examining such differences in literacy, a 2002 report used the joint availability of large-scale literacy assessments (the National Adult Literacy Survey (NALS) and the International Adult Literacy Survey (IALS) "to compare both the distributions and average literacy proficiencies of adults in the U.S. with those of adults in other high-income countries around the world" (Sum et al., 2002, p. 5). The report found that the average literacy score of U.S. adults was similar to the average scores of other countries, with the U.S. score falling in the middle of the range of scores from countries that were surveyed. More

[13]See http://www.pewinternet.org/2015/09/10/comparison-of-science-knowledge-questions-across-pew-research-center-surveys/ [July 2016].

[14]"Deficit perspective" is a term used to characterize viewpoints which focus on individuals' or population's weaknesses and which can at times associate differences between groups as weaknesses and unfairly characterize certain groups as deficient in some regard.

importantly, however, certain subgroups fell in the bottom half of the overall distribution, revealing a high degree of inequality between the best and poorest performers in the United States. The NALS results showed large gaps in the average scores of white adults and those of black and Hispanic adults. In addition, the mean literacy scores of young adults in the United States varied quite substantially by educational attainment.

These trends in foundational literacy are echoed in health literacy. In 2004, a report on the state of health literacy in the United States (Rudd et al., 2004) analyzed performance on 191 health-related tasks[15] in NALS and IALS. The results of this analysis highlighted significant inter-group differences, and included the worrisome finding that 12 percent of the U.S. adult population scored in the lowest bracket, with an additional 7 percent scoring at a slightly higher level that might still be expected to have great difficulty performing simple health-related tasks (Rudd et al., 2004, p. 3). Years of schooling, age, racial and ethnic status, and country of birth were all powerful predictors of health-related literacy skills among adults. Figure 3-2 shows the distribution of scores for each of these characteristics. The average overall score on this 500 point scale is 272 with a standard deviation of 61. In Figure 3-2(a), the average score of adults who had not completed high school or earned a general equivalency diploma (GED) (220) is lower than that for individuals who had graduated from high school or earned a GED (271) and lower than for those who had continued their education beyond high school (306). Figure 3-2(b) illustrates differences between racial/ethnic groups. Notably, among Hispanic adults, some 30 percent scored below Level 1. The report notes that several variables such as education, health status, poverty status, and immigrant status[16] influence the observed differences among racial/ethnic groups. Figure 3-2(c) shows that almost half of adults older than 65 in the United States performed in or below Level 1 (scores less than 25).

The Institute of Medicine (2004) delved deeper into the problem of health literacy, finding relationships between limited health literacy, poorer health status, and lower use of preventive services. These findings have had a considerable effect in the health care community, leading to a surge in research focused on interventions that could improve a patient's health literacy or on changes to the health care system that would reduce literacy demands on patients (e.g., Chin et al., 2007; Sudore and Schillinger, 2009). Some of this research is discussed in Chapter 5.

---

[15]For example, survey respondents were asked to read a medicine dosage chart and indicate the correct dose for a child of a particular weight and age, and to interpret information from a news article on bicycle safety (Rudd et al., 2004).

[16]The survey considered whether respondents were born in or outside the United States and found that 48 percent of Hispanic respondents were born outside the country.

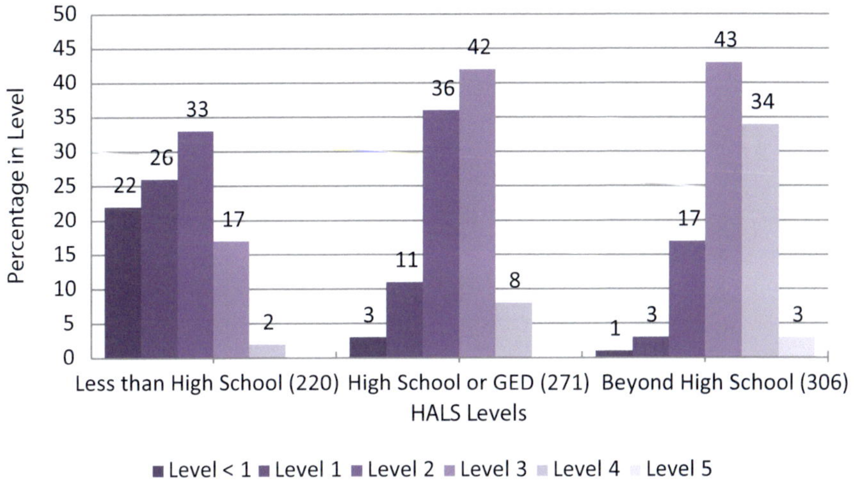

FIGURE 3-2a Average Health Activities Literacy Scale (HALS) proficiency and percentage at each level by education.
SOURCE: Rudd et al. (2004, Fig. 4).

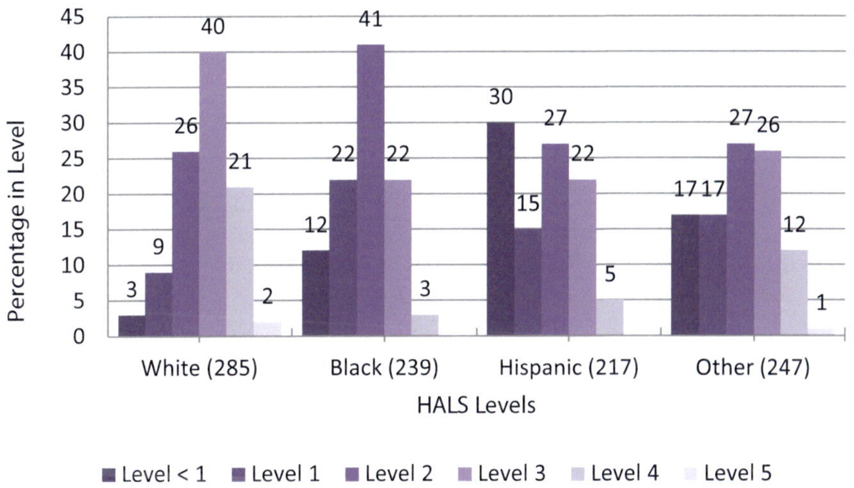

FIGURE 3-2b Average Health Activities Literacy Scale (HALS) proficiency and percentage at each level by race/ethnicity.
SOURCE: Rudd et al. (2004, Fig. 5).

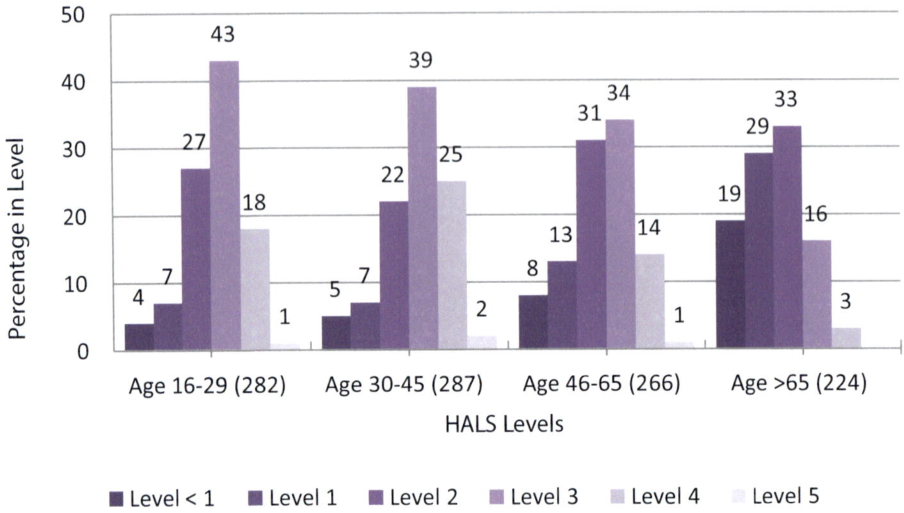

FIGURE 3-2c Average Health Activities Literacy Scale (HALS) proficiency and percentage at each level by age.
SOURCE: Rudd et al. (2004, Fig. 6).

THE ROLE OF SOCIAL STRUCTURES IN SHAPING SCIENCE LITERACY

Observed differences in science literacy (or at least in survey-based measures of science knowledge) lead quickly to more important questions: How do these differences arise, and what are their implications for society? Much focus has been placed on education and the school system. Differential access to high-quality science education is thought to have substantial impact on the ability of individuals to develop science literacy (Barton, 2003; Ladson-Billings, 2006). If educational opportunities are at least in part responsible for providing individuals with some of the features of science literacy identified in Chapter 2 (content knowledge, foundational literacy, understanding of scientific practices, for example), then if some individuals do not have access to high-quality education, they may be less likely to be able to develop the tools necessary for science literacy (see Box 3-3). Educational inequities also contribute to disparate levels of educational attainment and the under-representation of various groups in the science, technology, engineering, and mathematics (STEM) workforce, both of which may further reinforce the lack of knowledge or access to knowledge in a community (Anderson, 2015; Plutzer, 2013). But other systems of social and political life beyond schooling—including but not limited to the health

BOX 3-3
Differential Access in the U.S. Education System

In the United States, race and class remain two of the most pernicious indicators of access to educational opportunity. For example, the data on teacher preparation, considered an important indicator of instructional quality, show bleak disparities: students in high schools with higher concentrations of people of color or poor students are more likely to be taught science by a teacher without a major or minor in the subject (U.S. Department of Education, National Center for Education Statistics, 2004).

The data on course offering and course taking follow a similar pattern. Seventy-one percent of white high school students have access to the full range of math and science courses (algebra I, geometry, algebra II, calculus, biology, chemistry, physics) at their high schools, while only 57 percent of black high school students have comparable access (U.S. Department of Education, 2014a). Furthermore, 78 percent of the schools serving the lowest concentration of black and Latino students offer chemistry while only 66 percent of schools serving the highest concentration of black and Latino students provide the same opportunity. In 2009, only 65 percent of black high school students in the United States had taken chemistry, compared with 72 percent of white students and 85 percent of Asian students. Similarly, only 23 percent of students from the poorest high schools in the United States graduated having taken the "big three" high school science courses: biology, chemistry, and physics. From the wealthiest schools, 40 percent of students graduated having taken the big three (Center on Education Policy, 2007).

Given these inequities in access, it is not surprising that assessments of science knowledge show disparities in achievement. In 2011, 43 percent of white 8th-grade students were considered at or above proficient in science, compared with only 10 percent of black students and 16 percent of Latino students. Similarly, 45 percent of students whose families are not eligible for free or reduced lunch were considered at or above proficient, compared with only 16 percent of students who are eligible (National Center for Education Statistics, 2012b).

care system, the justice system, federal and state governments, and even informal learning networks—also shape how people interact with each other, with institutions, and with science information. All of these systems may operate differently across society as a result of class-based, regional, and/or cultural differences.

In all societies, life circumstances including (but not limited to) group membership and geographic location affect what people are likely to know. Southwell (2013) points to the isolating effects of segregated social networks that link some people, but not others, to resources, information, and expertise. Other researchers have pointed to research on the differential access and use of informal science education resources, nearly all of which indicates that underserved minorities, low-SES, and other marginalized groups are less likely to use these resources for a range of reasons including cost, distance, and the widespread (often accurate) sense that resources such as science museums were not designed with them in mind (e.g. Dawson, 2014). Also, despite a remarkable leveling in access to the Internet and Internet-enabled technologies in the past 2 decades in the United States, research continues to point to troubling differences in the way that people are supported in their use of Internet technologies, particularly in educational settings (Warschauer, 2006; Margolis et al., 2010; Barron et al., 2014). All of these differences feed into a Matthew effect "to those who have shall more be given" for science literacy (Merton, 1968; Stanovich, 1986).

Just as life circumstances shape what people are likely to know, life circumstances also shape what is important to know and the consequences befall those who are uninformed or unable (by virtue of personal characteristics or lack of access to resources) to interpret and act on available knowledge. For example, scholars working in the field of environmental justice have clearly demonstrated that environmental harms and goods are unevenly distributed in society (Brulle and Pellow, 2006). In the United States, poor citizens and people of color are far more likely to be exposed to hazardous levels of environmental toxins in their home and work environments, as well as in the air they breathe and the water they drink. In short, structural inequality produces situations in which some people have a more urgent and immediate need to engage with science around pressing health and environmental concerns—and those same people often have less access to informational and interpretive resources, as well as fewer avenues for political participation. This state of affairs leads the committee to think that the research question that could be asked is not what can these individuals do to better themselves, but what can society and scientific institutions do to help.

Much of the rhetoric around science literacy describes the obligations of citizens to participate in democratic governance. Although formal decision-making contexts like voting are the most obvious means of political participation, they are far from the only venues in which science may be relevant. Citizens (and noncitizen residents) may have access to political deliberation through open meetings; they may be able to offer input on new policies through comments on news sites, letters to their elected officials, community forums, citizen panels, and other public engagement mechanisms (Goidel and Nisbet, 2006; see Scheufele, 2011 for an overview and shortcomings of these mechanisms); and they may be able to exert influence through protest, petitions, referenda, and demonstrations (Jasanoff, 2003; Kleinman, 2000). Each form and instance of political participation may make different demands on science literacy, and many also provide formal or informal opportunities to learn more about science. When the public is expected or required to participate more, the demand for science literacy may be correspondingly greater, and opportunities for learning about socially relevant science may also be more common. In contrast, when there are relatively few avenues for political participation, or where access to the political process is restricted, the demand for science literacy (and the opportunity to develop and use it) may be correspondingly lower.

The pattern that is discernible in the formal and informal structures that shape political participation is also relevant to other social structures and sectors of society. In the justice system, for example, rising use as well as criticism of forensic science evidence (see, e.g., National Research Council, 2009) are increasing awareness of the need for science literacy among participants in court proceedings. Although this situation is hardly new, it is becoming increasingly difficult to ignore (Jasanoff, 2009). Rising awareness of the need for science literacy among citizens and professional jurists engaged in litigation serves as a call to the justice system to examine structures that promote and support science literacy in legal contexts and serve to provide relevant scientific information and training to judges, attorneys, and juries as necessary.

CONCLUSIONS

Currently, what is measured as science literacy at the society level comes from large public opinion surveys among adults and survey tests of adolescents in many countries. These measures are useful for examining trends over time and for identifying areas for improvement in a society. Additional data comes from tests of adolescents through PISA, which indicates that U.S. performance on recent tests of science knowledge is about average in comparison to a range of countries (Table 3-3). Again, the committee declines to draw conclusions about adolescents, as a thorough investigation and interpretation of the PISA data is beyond the scope of this report.

Indicators of adults' knowledge of science are limited to a narrow range of measures on public surveys that can be compared across countries. It is difficult to draw strong conclusions on cross-national performance from these measures. However, survey responses over time have shown much stability. In the United States, the average score on knowledge questions for each survey year since 2001 is consistently near 65 percent of the questions answered correctly. In addition to observed stability in responses, there is evidence that no country consistently outperforms other countries on all questions (Table 3-2). In view of these findings on adults' performance and the limitations of public opinion data, the committee concludes that the performance of the adult population in the United States on general science knowledge is similar to that of other countries for which there are data.

CONCLUSION 5 The population of adults in the United States performs comparably to adults in other economically developed countries on most current measures of science knowledge.

The committee recognizes that the narrow range of existing measures is unlikely to adequately capture the full range of science literacy discussed in this report. Although countries can be compared and ranked on the basis of their survey scores, there is little understanding of what those rankings mean: What do the differences in survey measures of science knowledge actually imply about the health and wellbeing of those societies, their communities, or their citizens? The current evidence available simply does not allow one to make strong inferences about the relationship between average scores on common measures of science literacy and the putative consequences of enhanced science literacy, such as improved public health or more enlightened civic and political engagement.

The large public opinion surveys in different countries also include measures of attitudes toward science. On these measures, there are many similarities among countries, and response trends have been stable across multiple survey years, particularly in the United States for which there are more data. Attitudes have been (and remain) quite positive, notably in regard to the perceived benefits created by science for societies and support for scientific research.

> **CONCLUSION 6** Current evidence, though limited, shows that populations around the world have positive attitudes toward science and support public funding for scientific research. These attitudes have been generally stable over time. In addition, the same evidence reveals an overall high level of trust in scientists and in scientific institutions.

In the cross-national comparisons, there is some evidence that countries with similar measures of economic development and educational attainment tend to have similar average scores on measures of science knowledge. Although this is useful and intuitive information, a focus on average levels can obscure the wide variation that can exist in a society. In reviewing the literature and data from surveys on science literacy as well as those on foundational literacy and health literacy, the committee found significant disparities in knowledge and access to knowledge. Much more is known about disparities in foundational literacy and health literacy as well as about disparities in science education in school systems in the United States. In contrast, research on stratification, variation, and disparities in science literacy is rare. The committee encourages new research in this area to examine the extent of disparities in science literacy and the social structures that contribute to inequities.

> **CONCLUSION 7** Within societies, evidence shows that severe disparities in both foundational literacy and health literacy exist and are associated with structural features such as distribution of income and access to high-quality schooling. Though direct evidence for such structural disparities in science literacy is scarce, we conclude they too exist, in part because

the possession of foundational literacy is so integral to the development of science literacy.

The committee proposes that any social structures that limit an individual's access to high quality educational opportunities would have a direct bearing on that individual's ability to develop or apply science literacy. Access to education is but one example. Studies of social structures reveal that groups disadvantaged in one social system tend to be disadvantaged in others as well, and the effects of disadvantage accumulate over generations, leading to harsh, self-reinforcing disparities (e.g., Ladson-Billings, 2006; Delgado and Stefancic, 2012).

Analysis of aggregated science knowledge measures could be used to analyze the variations in social structures as explanatory variables of the differences in national means and could be used to analyze subgroup differences in societies. However, a thorough account of disparities in science literacy cannot rely solely on aggregated measures of science knowledge. It would also include an examination of the conditions that structure science literacy differently for different groups of people, shaping how their understandings of science are developed and deployed, and the institutional response needed to address current failings. As discussed in the following chapters, different individuals and communities in a society rely on science literacy to different degrees at different times to accomplish their personal and civic goals. In this sense, social conditions define meaningful science literacy, and shape (if not determine) the distribution of science literacy for the communities and individuals therein. The research community has not yet studied in sufficient detail both the value of science literacy in societal systems (such as the health care system, the justice system, and the various systems of political participation) and the opportunities to *develop* science literacy that these systems provide.

4

Science Literacy for Communities

It is impossible to understand how science literacy affects a society without examining the emergence and importance of science literacy in the communities that comprise it. In Chapters 1 and 2 we outlined an overarching perspective in which science literacy emerges, and can usefully be studied, at multiple levels of social organization. In this chapter we expand on the idea that science literacy can be developed, possessed, and used by communities. The committee reviews the research on such communities with three guiding questions in mind: What are the characteristics of such communities? Do they display community-level science literacy? Is there evidence that they contribute to scientific knowledge and, if so, under what conditions?

We examine community-level science literacy through examples of communities that accomplish various goals, by virtue of their collective literacy, that cannot be easily attributed to the actions of any particular individual.[1] Science literacy in a community does not require each individual to attain a particular threshold of knowledge, skills, and abilities; rather, it is a matter of a community having sufficient shared resources that are distributed and organized in such a way that the varying abilities of community members work in concert to contribute to the community's overall well-being (e.g., Dewey, 1927; Roth and Lee, 2002; Ownby et al., 2014).

We define community broadly. Some communities are bound together by

[1]Health literacy researchers have also argued that successful community health outcomes rely on the collective effects of individuals' knowledge to foster improvement in health status for populations. Nussbaum (2000) argues that a key health goal for a community is to have people who have the capacity to firmly advocate for and to act in the best interest for the public good.

place or identity. Some could be described as "communities of practice," that is, groups of people whose mutual involvement have given rise, over time, to shared routines, activities, and goals, the value of which is understood in similar ways by all members (Lave and Wenger, 1991; Wenger, 2000). For the purposes of our analysis, communities are groups of people who are functionally interconnected in a way that enables exchange of information and are typically defined by shared goals and interests. However, we stress that many such communities are more loosely organized, more heterogeneous, and more transient than the archetypal community of practice.

It is important to note that communities need not be tied to particular geographic places. The examples of cities or localities as communities presented in this chapter reflect just one construction of a community; evolving communication technologies allow people from all over the world to exchange information and to connect with others on issues of interest and relevance to them (Brossard and Scheufele, 2013). The pervasiveness of Wikipedia and social media platforms, for example, encourages the formation of communities that span geographic boundaries.

The committee has also chosen, in this chapter, to treat science literacy and health literacy together. In particular cases, a community's experience of health and illness often drives engagement with health systems, which leads to questions about the underlying science: if and when those questions are answered, the answers are often applied to the same set of health systems. It is conceivable that a community's capacity to interact with health systems is concentrated in certain individuals or subgroups in the community, while its capacity to understand and contribute to science is concentrated in other individuals or subgroups, but, to date, no data specifically address this claim.

ORGANIZING AROUND SCIENCE AND HEALTH ISSUES

In his groundbreaking work on AIDS treatment activism, Steven Epstein (1995) offered an account of how a group, most of whose members (initially) possessed little formal knowledge of science, went about developing expertise in an emerging and contested research field. Alarmed at the rapid progress of the disease and high mortality rates, AIDS activists developed scientific knowledge (including an understanding of scientific practices) to demand modifications to drug-testing procedures and to the Food and Drug Administration's drug approval policies; working together, they successfully advocated for alternatives to the placebo-control protocol for clinical trials (previously a cornerstone of biomedical procedures) in order to expedite the delivery of drugs to consumers in what can be characterized as health emergencies.

The activist, science-savvy, and ultimately influential community that organized itself around AIDS treatment and research had many advantages that no doubt contributed to its success. Epstein (1995, p. 415) notes that "the gay community had pre-existing organizations that could mobilize to meet a new threat, and it mattered that these communities contained (and in fact were dominated by) white, middle-class men with a degree of political clout and fund-raising capacity unusual for an oppressed group." The community was able to acquire credibility in certain domains of scientific practice because biomedicine (in comparison with other areas of science and technology) was relatively open to outside scrutiny, and participation in clinical trials gave members of the community a unique and valued perspective on the treatment process. That is, the community was socially positioned to challenge the status quo and had unique access to the particular realm they hoped to affect.

The AIDS activism documented by Epstein is just one notable example of a social movement for which some science literacy was crucial. Social movement research is a well-established field that has recently turned its attention to the intersection of science, technology, and health with social and political advocacy. This field is now home to an increasing number of research programs that are studying science literacy at the community level.

A related phenomenon is popular epidemiology, first defined in the early 1990s by sociologists studying communities with unusually high incidence of cancer linked to industrial pollution (Brown, 1992; Allen, 2003). Phil Brown, the researcher who coined the term, describes popular epidemiology as community-based activists working together to "detect and act on environmental hazards and diseases," a process that requires them to "gather scientific data and marshal the knowledge and resources of experts" (Brown, 1993, p. 18).

In Brown's first well-known account of popular epidemiology, a group of families in Woburn, Massachusetts, "confirmed the existence of a leukemia cluster and linked it to industrial chemicals that were infiltrating their public water supply" (Brown, 1993, p. 17). Using ethnographic data, Brown describes how the Woburn families identified a problem in their community (high rates of cancer) and used their collective resources to develop scientific knowledge (including an understanding of scientific practices and judging appropriate scientific expertise) to address that problem.

Not all studies of community-based science literacy focus on health topics. In participatory environmental monitoring (also called "community-based monitoring") community groups collect data to monitor environmental systems. Kinchy and colleagues (2014, p. 260) note that "from the venerable Audubon Christmas Bird Count to the recent phenomenon of 'hacker' or DIY water testing kits, diverse publics are engaging in work to monitor and measure changes to the environments in which they live." Hundreds of thousands of such community groups have formed around the world (Pretty, 2003; Conrad and Hilchey, 2011). In New York and Pennsylvania alone, dozens of community groups are engaged in monitoring the effects of "fracking" on local watersheds

(Kinchy et al., 2014).[2] In many parts of the world, community coalitions have used relatively simple, homemade air quality monitoring devices to produce surprisingly reliable measurements and hold industrial and agricultural polluters to account (Conrad and Hilchey, 2011; O'Rourke and Macey, 2003). Although such work is not new, it is becoming increasingly common and has increasingly been the subject of study over the past two decades (Conrad and Hilchey, 2011).

These bodies of research reveal that intensive community-level engagement with science is surprisingly common; taken together, they provide an empirical foundation for conceptualizing science literacy at the community level.

SCIENCE-LITERATE COMMUNITIES

Although the above examples are compelling in their own right, the question that concerns us here is to what extent the communities formed to pursue social activism, popular epidemiology, or participatory environmental monitoring can be characterized as science literate. The evidence is compelling that the success of these communities in promoting policy changes and other outcomes depends, at least in part, on their ability to develop knowledge of science- and health-related issues, as well as knowledge of general scientific practices, and on their capacity for sophisticated interaction (both internally and externally) with scientists and health professionals, scientific institutions, and health systems. For example, Kaplan (2000, p. 75) reports that activists in communities around the Hanford nuclear facility in Washington "became well educated on science and technical issues of radiation health effects, Hanford operations, and nuclear waste storage." Epstein (1995, pp. 417-418) offered a compelling account of how AIDS activists developed knowledge in support of their goals:

> Activists often begin with the examination of a specific research protocol in which patients have been asked to participate and, from there, go on to educate themselves about the mechanism of drug action, the relevant "basic science" knowledge base (such as considerations of the viral replication cycle of HIV or the immunopathogenesis of AIDS), and the inner workings of "the system" of drug testing and regulation including the roles of the pharmaceutical companies and the relevant government advisory committees.

Some community groups replicate particular techniques and practices that professional scientists use in their work. Kinchy and colleagues (2014, p. 278) note that "civil society organizations in the field are typically aware of the scientific standards that academic and regulatory scientists use in their analysis of water quality, and many seek to align their own monitoring practices with those of recognized experts in the field" (see also Roth and Lee 2002; Roth and

---

[2]Hydraulic fracturing, or "fracking," is the process of drilling and injecting fluid into the ground at a high pressure in order to fracture shale rocks to release natural gas inside.

Calabrese Barton, 2004; Roth and Lee, 2004). In general, accounts of the learning that takes place in these community contexts are quite common, though they typically lack sufficient detail to reveal individual and collective learning processes (see, e.g., Evans et al., 2005).

Knowledge of science is often perceived by community groups as an important tool in pursuit of their goals. Brown (1993, p. 20) reports that ". . .Woburn activists found pride in learning science, a way to protect and serve their community, a means of guaranteeing democratic processes, and fueling of personal empowerment." Kaplan (2000, p. 81) describes how activists around Hanford "demonstrated the ability to take an active role in deciding what science and technology policies pose a danger to public health and the environment and the ability to work together to change those policies."

Yet in the context of community groups, it is important to recognize that scientific knowledge is not held apart from other sorts of knowledge as something special and different. Instead, it is treated as part of the story, essential information that must still be "reconstructed" and understood in proper context before it becomes useful (Layton et al., 1993; Irwin and Wynne, 1996). Often, scientific knowledge is juxtaposed with local experience and knowledge from other expert domains, such as law or economics, to produce "interestingly hybrid . . . ways of knowing and varieties of expertise" (Epstein, 2008, p. 518; see also Hess et al., 2008).

One set of circumstances that deserves special consideration is the juxtaposition of scientific and community knowledge that arises when indigenous communities interact with science. In the past 2 decades an expanding body of research in both the social sciences (e.g., Ross et al., 2007) and the natural sciences (e.g., Huntington, 2000) has explored how indigenous and scientific knowledge[3] can be integrated, often with positive consequences for both community-based concerns and scientific understanding (Drew, 2005). Successful collaboration appears to depend (among other things) on both the acceptance of scientific and indigenous expertise as dynamic ways of knowing rather than static bodies of information (Berkes, 2009) and the presence of people capable of acting as intercultural "knowledge bridgers" (Bohensky and Maru, 2011).

In each community the distribution of labor takes various forms. Some, though not all, of the existing research literature points to the importance of strong individual leaders who aid in the coordination of knowledge and resources scattered throughout the community. Existing networks and organi-

[3]Some scholars argue that the divide between indigenous knowledge and science is artificial and that indigenous knowledge systems should be considered scientific in their own right (Agrawal, 1995; Bang and Medin, 2010). Others argue that using "science" as an umbrella term that includes indigenous knowledge systems obscures how the fields that are typically referred to as science evolved in and reflect the norms and biases of specific places and cultures (Turnbull, 1997, 2003).

zations can also provide a useful structure. Churches and clergy have played a significant role in communities facing environmental struggles (see, e.g., Brown, 1993). The AIDS treatment activism movement, described above, brought together a broad and diverse group of community members, including grassroots activists and advocacy organizations to health educators, journalists, writers, service providers, and people with AIDS or HIV infection (Epstein, 1995, p. 413). The connections among diverse stakeholders, whether they had already existed or were formed in response to a particular issue, enable people in a community to benefit from each other's varying knowledge and influence and are therefore critical in positioning individuals and organizations to act collectively (Lee and Roth, 2003). More research is needed to understand the relationship between network structure and community-level science literacy.

It is important to note, however, that once mobilized, communities are not necessarily united in their goals. There are often multiple organizations and actors addressing a particular issue, as well as divisions about goals (Roberts and Toffolon-Weiss, 2001). Grand Bois, Louisiana, offers a rare example of total community cohesion in which all 301 residents of the community joined a class-action lawsuit against Exxon for dumping oilfield waste in open pits. Residents discovered the problem, conducted intensive popular investigation, and formed strategic alliances with university scientists and state legislators (Roberts and Toffolon-Weiss, 2001).

Although laypeople working in social movements and advocacy groups can develop and possess impressive scientific competence in relation to their particular concerns, many such communities include scientists or work directly with scientists and scientific groups to achieve their goals. The participation of scientists and other technical experts can be important to achieving a community's goals and establishing credibility (Shirk et al., 2012). For example, Phadke (2005) describes how engineers were crucial resources in an Indian community's efforts to redesign a dam and avoid forced relocation. In the AIDS treatment activism movement in the United States, some key players were themselves doctors, scientists, and nurses and were therefore able to facilitate communication between experts and laypeople (Epstein, 1995).

Scholars have classified various sorts of scientist-community partnerships according to who instigated the partnership, who holds power over action strategies and goals, and how deeply nonscientists are involved in scientific work (Moore, 2006; Bucchi and Neresini, 2008). In at least some cases, community groups are responsible for identifying and recruiting relevant experts. In popular epidemiology, for example, Brown (1993, p. 21) notes that community activists typically "find their own experts." Finding an appropriate expert is only the start, however, and forging useful collaboration is not always easy: Hess and colleagues (2008, p. 487), reviewing the literature on science and social movements, observe that "social movements, scientists, and entrepreneurs are uneasy allies and partners, and alliances sometimes shift into conflict and hostility."

Scientists who wish to become involved in such politically charged fields as environmental justice sometimes do so with caution and even secrecy—performing crucial technical work, such as analyzing samples, as well as important epistemological work, such as identifying key sources—while avoiding any public acknowledgement (Frickel et al., 2015). Overall, managing mutually satisfactory connections with scientists and scientific organizations may be an indicator of a community's scientific sophistication.

Taken together, evidence from case studies suggests that the success of these communities depends, at least in part, on the development of scientific knowledge throughout the community and the community's organization and composition, including the strength and diversity of connections with scientists and health professionals, scientific institutions, and health systems. The existing research shows that these community groups take many different forms and are widely variable in their duration and impact, but at least some have had an impressive and well-documented impact (see, e.g., Hess et al., 2008; Conrad and Hilchey, 2011).

PRODUCING SCIENCE KNOWLEDGE

While pursuing their own ends, communities may meaningfully contribute to new science knowledge. Although science literacy does not require making an original contribution to scientific knowledge, the committee asserts that the creation of new scientific knowledge is a compelling demonstration of science literacy. Research from the fields of health and environmental social movements, participatory environmental monitoring, and popular epidemiology converges on the finding that communities can and do contribute to new scientific knowledge in diverse and substantive ways (see, e.g., Kinchy et al., 2014; Bonney et al., 2009; Cohn, 2008).

Although health advocacy communities and movements may primarily be concerned with promoting health rather than contributing to the creation of new knowledge, they can nonetheless make epistemic contributions: Hess and colleagues (2008, p. 481) observe that health social movements "push the boundaries of science in new directions and challenge identities and interests on both sides of the lay-expert divide." Conversely, for many community environmental groups, creating new knowledge is an acknowledged part of their mission: Kinchy and colleagues (2014, pp. 275-276) report that the large majority of community organizations in a study of participatory environmental monitoring reported that "one of their objectives is to contribute to scientific knowledge" (see also Bonney et al., 2014). Contributions from these groups, and others like them, are valuable to both scientists and the communities themselves, and they constitute an important consequence of science literacy at the community level.

In education and natural science research, the best-known examples of community groups participating in scientific research (and thus contributing to the creation of scientific knowledge) are those in which people serve as volunteer data collectors in large-scale scientific projects. Although these projects are of well-established value for scientists (see Bonney et al., 2009; Cohn, 2008), they do not always require much in the way of scientific knowledge from their participants, who in many cases are involved in focused data collection tasks that by themselves generate little learning about science (e.g., Brossard et al., 2005). Lakshminarayanan (2007) makes a valuable distinction between projects that position community members as scientists (or, more generally, contributors to knowledge) and projects that use people for science—positioning them as tools or instruments capable of collecting data but little more. Most of the examples discussed in this chapter fall into the former category, in that non-scientists in communities are involved in posing as well as answering questions and interpreting as well as collecting data. Brown (1993, p. 39) notes that community groups involved in lay epidemiology "may initiate action and even direct the formulation of hypotheses." Epstein (1996) describes in detail how AIDS treatment activists transformed the clinical trial process to be more responsive to the needs and interests of patients and research participants. More broadly, Frickel and colleagues (2010, p. 462) demonstrate that social movement organizations can shape research agendas in positive and negative ways—affecting both what science is done and what science "ought to remain undone." Though the production of new knowledge was not the goal for most of these communities, in pursuing their goals they also produced new scientific knowledge.

In the health literacy research community, this type of work is often conducted and discussed as community-based participatory research (Israel et al., 2005; Minkler et al., 2008). This approach focuses on building relationships—with principles of co-learning, mutual benefit, and long-term commitment—between scientists and community partners and incorporates community theories, participation, and practices in research efforts (Wallerstein and Duran, 2006). Israel and colleagues (2005) examined community-based participatory research in New York, California, Oklahoma, and North Carolina, documenting partnerships that researched environmental health problems and worked to educate legislators and promote relevant public policy. At each of the sites they studied, the pooling of diverse skills, mutual respect for the expertise of other partners, and a co-learning environment in which additional skill building took place contributed to community capacity building and partnership development (Israel et al., 2005, p. 1470). In Detroit, Michigan, the East Side Village Health Worker Partnership was created to examine and address social determinants of health. Through a series of group discussions and in-depth interviews, people in the community were instrumental in identifying key variables to examine, selecting and modifying measures to be included in a survey questionnaire, interpreting results, and applying findings to guide interventions. In this case, community participation in data analysis and interpretation strengthened community capacity and provided unique insights, contributing to the creation of scientific knowledge (Cashman et al., 2008).

The value of community participation in scientific research is widely recognized and supported by evidence. Community involvement can bring new questions to light, provide data that would otherwise be unavailable, encourage the integration of qualitative and observational data with experimental data, increase the robustness and public relevance of data collection strategies, garner political and community support for conclusions, produce new instruments and technologies, and build community awareness and knowledge (Allen, 2003; Clapp, 2002; Israel et al., 2005; Epstein, 1996; Hess et al., 2008; Conrad and Hilchey, 2011). Interestingly, the hybrid nature of community-scientist collaborations may be an important part of their strength (see, e.g., Bäckstrand, 2003; Funtowicz and Ravetz, 1995; Bucchi and Neresini, 2008; Corburn, 2007). Hess and colleagues (2008, p. 484) write:

> As activists and environmental professionals work together, many have become convinced of the need for heterogeneity in environmental problem-solving models. By recognizing the different bases of lay and scientific knowledges, activists and scientists may develop deliberative processes that allow for synergy between lay and expert knowledges.

Of course, not all communities organized around issues of health and science fully incorporate science knowledge into their work or generate new knowledge. For example, vaccine hesitant individuals can form communities, either within geographic regions or on social media (Cooper-Robbins et al., 2010; Dunn et al., 2015). These groups produce content that is distributed through a wide range of platforms, including social media applications such as YouTube and Twitter (Dunn et al., 2015). Though these communities are, from their own perspective, working to achieve a particular goal on a science-related issue, they interpret particular scientific arguments and findings in ways that diverge from the scientific consensus, and they do not build or apply science knowledge in the ways that many in the scientific community (and the public health community) would prefer. In this case, as in all communities, a community's behaviors and attitudes toward science are affected by that community's political and religious ideologies, which in particular circumstances may override scientific knowledge as an influence on dispositions and actions (Fiske and Taylor, 1991).

CONSTRAINTS ON COMMUNITIES

Although many communities suffer environmental or health crises, research does not yet show the extent to which communities are able to mobilize to respond to these problems at a local level or what features of particular communities enable them to develop and use science literacy in powerful ways. Science literacy at the community level, like science literacy at the individual or society level, is to some extent a product of the larger social structures in which a community operates. As noted above, the success of AIDS treatment activism was possible in part because biomedicine as a field was relatively open to outside scrutiny and because the activists were themselves participants in the contested studies. Furthermore, and crucially, their ability to accomplish what they did was shaped by the structural privilege (such as the race, gender, and/or class status) belonging to some activists. That is, this particular community was socially positioned to challenge the status quo and had a relatively high degree of access to the particular social system they hoped to change.

Other communities may be hampered by their inability to establish legitimacy and credibility (Ottinger, 2010, 2013; Ottinger and Cohen, 2011). Ottinger (2010) shows how regulatory standards and standardized practices cemented resistance to citizens' broader participation in air quality monitoring in Norco, Louisiana. Activists promoted an alternative to standardized air monitoring practices, using methods that measured short-term spikes in air pollution levels rather than the standard strategies that measure the average concentrations of toxic chemicals over long periods. Regulatory standards for air quality, combined with standardized practices for monitoring, meant that activists' methods and data were incompatible with the standard scientific practice. In this case, existing standards provided grounds for excluding nonscientists from decision making—not because they were not experts, but because they did not have "credible" scientific information to offer (Ottinger, 2010).

Access to knowledge or particular resources is also critical. Greenberg and Wartenberg (1991) report that many state and local health departments, responding to citizens' reports of high rates of cancer in their communities, have sent out form letters declining to investigate claims of potential cancer clusters. Lack of open access to data, from various sources, including government agencies, may hamper a community's ability to build or apply its science literacy.[4]

Under-resourced communities are particularly susceptible to the types of crises in which community activism informed by science literacy would be crucial, yet they often have the least access to resources that support science literacy. For people who rely on municipal water supply in Flint, Michigan (Washington and Pellow, 2016), or who live and work in the part of Louisiana known as cancer alley (Allen, 2003), or in the agricultural towns of the central valley in California (Harrison, 2011), there is an urgent need to understand the perils of environmental health threats (see also Fessenden-Raden et al., 1987). In these contexts, developing and using science literacy may play an important role in protecting one's family and community and in advocating for social and environmental change. Yet these and other communities most affected

[4]In some cases, individuals and communities have difficulty accessing their own data from community-based participatory research groups. This can create a significant barrier to community activism around knowledge generated by those groups.

by environmental harms are often the same communities that are structurally disadvantaged in both the development and use of science literacy, in the ways described in Chapter 3. They are also subject to economic, social, and political pressures that constrain their ability to act on what they know.

Given that the ability of certain communities to build and apply their science literacy is constrained by social structures, developing science literacy in communities may also require supporting and empowering communities to act on knowledge. For example, research from the health literacy field suggests that building health literacy requires a broad range of educational and communication methods (such as personal forms of communication and community-based educational outreach), as well as service management and organizational supports (such as minimizing and simplifying form filling) (Nutbeam, 2008).

The shifting contexts and social structures in which communities operate play an important role in determining why one community may build and apply science literacy while others do not. Additional research is needed to understand the various features (e.g., community organization) and contexts (e.g., a community's political power in a particular setting) that enable or prevent community involvement and action related to science literacy. More specifically, carefully constructed quasi-experimental designs comparing communities that have experienced similar[5] issues might help identify the community features that lead to the development of science literacy at the community level (see, e.g., Campbell and Stanley, 1963; Trochim, 2000). For example, studies could consider: How is knowledge sought and shared with a community? How easy is it to access the scientific resources of a community? What aspects of a community's culture are most important in shaping its engagement with science? (Kickbush, 2001).[6]

CONCLUSIONS

Existing research provides compelling support for the idea that communities can possess and use science literacy to achieve their goals and may also contribute to new science knowledge in doing so. The ability of communities to apply their science literacy is enabled or constrained by social structural contexts. Scholars in the health literacy field recognize that improving health literacy in a community involves more than the exchange of information; the research indicates that developing science literacy in communities may also require supporting (through institutional systems) and empowering collective

[5]Care would need to be taken in the selection of studies for comparison, as issues that appear superficially similar to outside observers may not be analogous. The historical context of particular incidents shapes the events that follow (see, e.g., Wynne, 1992).

[6]Kickbusch (2001) recommended that a health literacy index be developed to reflect the composite health competence and capabilities of a community as it relates to a set of health, social, and economic outcomes.

groups of individuals—communities—to act on their knowledge (Nutbeam, 2008).

CONCLUSION 8 There is evidence from numerous case studies that communities can develop and use science literacy to achieve their goals. Science literacy can be expressed in a collective manner when the knowledge and skills possessed by particular individuals are leveraged alongside the knowledge and skills of others in a given community.

CONCLUSION 9 Based on evidence from a limited but expanding number of cases, communities can meaningfully contribute to science knowledge through engagement in community action, often in collaboration with scientists.

Most of the case studies the committee examined to explore community-level science literacy did not focus explicitly on science literacy or health literacy. Rather, we extrapolated the role of science literacy in accomplishing a community's goals. Future research should explicitly consider the development and uses of science literacy in community contexts and its value in achieving community goals.[7] Furthermore, in most cases there has been little effort to assess, prior to the start of community action or controversy, what level of science literacy was present in the community at the level of individuals in the community or how it is distributed.[8] It is difficult to imagine doing so in most cases, since community action and controversy are typically what draws the attention of researchers. As a result, it may be difficult to test a cause-and-effect relationship between "enhanced" science or health literacy and particular community-level outcomes, so that one has to infer what a community had to know in order to accomplish a particular goal.

Though the committee supports a view of science literacy that considers how communities may possess science literacy, we also believe that assessing the nature and outcomes of science literacy at the community level should be approached with caution. When a community achieves something impressive, such as preventing the construction of a hazardous waste storage facility in its area, there is no guarantee that it has done so primarily through the application of some form of science literacy. Therefore, the research on popular epidemiology, environmental and health social movements, and community environmental monitoring is especially valuable: in each case, researchers have examined how and when communities possess knowledge about science and exert influence on the creation of new knowledge.

[7]For the committee's specific research recommendations, see Chapter 6.

[8]Fessenden-Raden and colleagues (1987) consider inter- and intracommunity differences in cases of risk perception communications.

5

Science Literacy for Individuals

The nature of an individual's science literacy is tied to the social organizations in which individuals function. System-level factors, such as the differential distribution of knowledge and resources, affect the circumstances that individuals or communities confront, which demand or promote the application of science literacy. These social structures (described in Chapter 3) include a society's or a community's education system, health care system, justice system, and governance structure. Results from aggregated individual-based science-knowledge assessments in the United States show differences by race, ethnicity, age, gender, and educational attainment, which are in no small part based on the differing constraints and opportunities that face these subpopulations (see Chapter 3).

This chapter focuses on the value of science literacy and health literacy for individuals. We assess the current empirical evidence regarding the outcomes of science literacy for individuals, including evidence on the association between science literacy and attitudes towards, perceptions of, and support for science and the relationship between science literacy, health literacy, and behaviors (particularly behaviors related to health).

THE RELATIONSHIP BETWEEN SCIENCE LITERACY AND ATTITUDES

As discussed in Chapter 2, science literacy at the individual level has largely been measured by assessing an individual's knowledge level using content knowledge assessments and measures of understanding of scientific principles.

The items measured are not assumed to be exhaustive of the concept in question; rather, they form a sample from a larger set of potential items that could have been chosen to represent the unobserved characteristic, in this case, science literacy. Individuals who know the answers to particular textbook-style questions about biology, physics, and chemistry have a high probability of knowing other facts in the same domains. In the same way that the questions that graduate school candidates face in the GRE General Test are not the things that the test takers directly need to know for their graduate work but are indicative of necessary and related abilities and knowledge, content knowledge assessments (particularly the Oxford scale science literacy items) are designed to distinguish generally among people who have relatively more or less science knowledge.

The extent to which these knowledge measures have validity in measuring levels of scientific literacy is discussed in Chapter 2, and many scholars, as well as this committee, recognize that they may capture only a narrow aspect of science literacy. That is, they do not specifically assess the many other features of science literacy (detailed in Chapter 2). However, these knowledge measures are the ones most commonly used in studies exploring the potential relationship between science literacy and attitudes toward, public perceptions of, and support for science, as measured by the public's evaluation of the social impact of science and technology.[1]

This section discusses the major scholarly research that considers the potential link between science knowledge and attitudes toward and public perceptions of science, with special attention to the association between science knowledge and support for scientific funding; see Box 5-1. It explores research examining the relationship between science knowledge and a set of broad attitudes toward science that reflect an individual's assessment of the scientific research enterprise generally, as well as the relationship between knowledge and a more focused set of attitudes toward specific scientific controversies (such as nuclear power, climate change, stem cell research, and genetically modified foods). In addition, it discusses potential moderators and mediators to this relationship and recent experiments aimed at testing interventions to increase science knowledge and their impact on an individual's attitudes toward science.

The committee recognizes that this may not be the only important empirical question. There is a large body of literature on the influences of attitude

---

[1]The wordings of many attitudinal items, which are typically worded positively, may be subject to acquiescent response bias (Bauer et al., 2000). Given these measurement concerns, Bauer and colleagues (2000) proposed an alternative, multi-item measure of public science attitudes that focused on understanding the nature of science (factual knowledge, methodological knowledge, and knowledge of the scientific institution). This measure of science attitudes is primarily concerned with understanding people's views on the controversy over the nature of modern science. Despite cross-validating their measure with multiple samples, the assessment is not currently widely used in research; it presents an opportunity for future research.

BOX 5-1
The Relationship Between Science Knowledge and Support for Scientific Funding

Examining data from the United States, Besley (2016) found that bivariate relationships exist between support for scientific funding and demographic characteristics (gender, race, and education), use of science communication channels (including museum visits and Internet and newspaper use), science knowledge, and attitudes about science and scientists. Knowledge measures had a very small, positive relationship with funding, explaining only 10 percent of the variability in support, indicating that other variables most likely have more explanatory power. Studies from a range of other countries have found an association between science knowledge and support for funding (Muñoz et al., 2012; Sanz-Menéndez and Van Ryzin, 2013; Sanz-Menéndez et al., 2014).

Scholars have also examined the relationship between scientific knowledge and support for funding of specific science. Liang and colleagues (2015), for example, found that factual knowledge of nanotechnology was positively and significantly related to support for nanotechnology funding in Singapore, while perceived familiarity with the technology had a positive and significant effect in both the United States and Singapore. Conversely, Scheufele and Lewenstein (2005) found that, after controlling for demographic predispositions and information seeking in mass media, knowledge about nanotechnology was largely unrelated to attitudes toward increased funding for nanotechnology.

Recent work includes evidence that science knowledge is a small predictor of positive views about the role of science in policy and support for science funding on its own. However, these studies also show that the relationship is likely moderated by ideology and religiosity (Brossard et al., 2009; Ho et al., 2010; Fung et al., 2014; Carl and Cofnas, 2016; Gauchat, 2015). Practically, this means that while science knowledge can be a reasonable predictor of support for science funding, the relationship varies. For example, Gauchat (2015, p. 739) found that left-right or liberal-conservative political orientation is a factor in predicting support for science funding among more scientifically sophisticated respondents.

formation, such as value predispositions, media use, and perceptions of risks and benefits, among others; see Box 5-2. Though this research literature is informative, the statement of task guided the committee's narrow exploration of the relationship between science knowledge and attitudes.

Science Knowledge and Attitudes Toward Science

Among the most heavily cited analyses assessing the direct link between science knowledge and attitudes toward and public perceptions of science is a meta-analysis that analyzed publicly available survey data from 193 surveys conducted across 40 countries between 1989 and 2004 (Allum et al., 2008). The

BOX 5-2
Factors Other Than Knowledge That Influence Attitudes Toward Science

The debate surrounding the exact role that knowledge plays in attitudes toward, public perceptions of, and support for science has been at the core of discussions about science literacy and public understanding of science. Some scholars have criticized approaches that seek to improve public science attitudes by increasing an individual's scientific knowledge. They argue that this "deficit model" (Wynne, 1992) unfairly characterizes lay people as "deficient" relative to scientific experts (Sturgis and Allum, 2004). Others argue that models that privilege knowledge in the opinion formation process are overly simplistic (see, e.g., Besley, 2010; Besley and Oh, 2014; McComas et al., 2014; Brossard et al., 2005; Brossard and Nisbet, 2007; Cacciatore et al., 2012a; Cacciatore et al., 2011; Cacciatore et al., 2012b; Lee et al., 2005; Nisbet, 2005; Scheufele and Lewenstein, 2005; Scheufele, 2006; Fiske and Taylor, 1991). More specifically, these scholars have highlighted that individual factors other than knowledge can have a significant influence on attitudes toward science. The committee has chosen to highlight a select number of these factors: media use, value predispositions, and trust.

Media Use Scholars have focused on the role of behaviors in shaping attitudes toward science, particularly the relationship between an individual's media use and attitudes. Findings show a relationship between media use and attitudes toward science in general (e.g., Nisbet et al., 2002; Dudo et al., 2010), as well as specific science issues (e.g., Scheufele and Lewenstein, 2005; Brossard and Nisbet, 2007; Ho et al., 2008). For example, Ho and colleagues (2008), in an analysis of media use and public attitudes toward embryonic stem cell research, found that attitudes toward stem cell research were shaped by cues from the news media. These results suggest that mass media provide an important part of the social context by which citizens perceive controversial science.

meta-analysis only examined factual knowledge items because items relating to the scientific method and understandings of science were not found in all of the studies considered. The measured attitude items can be classified into five areas: general science attitudes and specific attitudes toward nuclear power, genetic medicine, genetically modified foods, and environmental science.

Controlling for measures of age, gender, and education that were common to all 193 datasets, the meta-analysis found that there was a small, positive overall relationship between science knowledge and attitudes. Equally important, however, the study found that the size of this relationship varied substantially by whether the measure of attitudes was focused on general science or a specific topic and whether the knowledge measure was a general science measure or one

Value Predispositions Value predispositions, such as political ideology, religiosity, and deference to scientific authority have been shown to affect attitudes toward science. In a study of perceptions of embryonic stem cell research, Nisbet and Goidel (2007) found that value predispositions related to Christian conservatism and social ideology influenced citizen evaluations about that research. Brossard and Nisbet (2007) found a direct and positive relationship between deference to scientific authority and support for agricultural biotechnology. Strength of religious beliefs was found to be negatively related to support for funding of nanotechnology (Brossard et al., 2009).

Trust A large literature also examines trust, defined as having multiple dimensions, including integrity, dependability, and confidence (National Research Council, 2015). In this literature, scholars have shown that trust in scientists and scientific institutions affects attitudes toward science (Sjøberg, 2002) as well as attitudes toward specific science issues. For instance, Priest and colleagues (2003) explored public perceptions toward biotechnology and found that "trust gaps" (i.e., the size of the difference in trust in different stakeholders involved in the technology) emerged as a predictor of biotechnology attitudes. Research also suggests an inverse relationship: attitudes may influence trust. Roduta Roberts and colleagues (2013) found that attitudes, rather than perceived knowledge, led directly to an increase in trust in science and technology.

This does not mean that science knowledge levels do not matter, but, rather, that they may be affecting individuals differently depending on their values and other factors. Brossard and colleagues (2009) provided initial evidence for this type of indirect relationship in their study of nanotechnology. Religiosity served as an interpretive tool for individuals to make sense of nanotechnology. That is, levels of knowledge interacted with religiosity such that the link between knowledge and support was significantly weaker for highly religious respondents than it was for less religious respondents.

focused on a specific type of knowledge. Specifically, the correlation between general scientific knowledge and a range of specific science attitudes was generally weaker than the correlation between general scientific knowledge and general scientific attitudes. For example, the data suggested almost no relationship between general science knowledge and attitudes about genetically modified food, a potentially negative relationship between biology-specific knowledge and attitudes about genetically modified food, and a small, but negative relationship between that same general science knowledge measure and attitudes toward environmental science (see also Gaskell et al., 2004; Priest et al., 2003). The results further suggested that the basic relationship between general science knowledge and general attitudes was slightly larger than initially estimated but still small, with little cross-national variation.[2] Ultimately, the authors characterized the relationship in the following way (Allum et al., 2008, p. 51):

> Our findings suggest that, if one examines all measured knowledge and attitude domains, there is a small but positive relationship. Perhaps we might characterize the importance of this as "shallow but broad." Those scholars who take the falsity of the "deficit model" as axiomatic will no doubt want to focus on the low magnitude of the overall effect. Those who believe that "knowledge matters" will likely emphasize the robustness of the relationship—over so many national contexts and over time.

Other scholars have reinforced these findings and shown that this relationship becomes more complicated when assessing specific science knowledge and attitudes. Individuals may have broadly positive (or negative) attitudes toward science and may hold a set of attitudes toward specific scientific issues or disputes that do not align with their general attitudes toward science. For example, O'Connor and colleagues (1999) found a positive relationship between specific attitudes (willingness to take voluntary actions to address climate change) and specific knowledge (understanding of climate change). The study found that willingness was positively related to knowledge of the causes of climate change, although the relationship was weaker once measures of overall environmental attitudes were included in the analysis. In contrast, Bauer and colleagues (1997) analyzed three separate years of Eurobarometer data and found that, while specific knowledge of biology increased across the three surveys, optimism about both biotechnology and genetic engineering actually decreased during that time. The authors also found that scientific knowledge was only weakly correlated with a host of application areas for either biotechnology or genetic engineering. Other studies have concluded that higher levels of scientific knowledge were correlated with negative perceptions of biotechnology (e.g., Midden et al., 2002), pointing to the inconsistent results across studies trying to assess a direct relationship between knowledge and attitudes.

Priest and colleagues (2003) explored public perceptions toward biotechnology in both the United States and Europe. While the authors found science knowledge and educational levels to correlate differently with several different application areas of biotechnology (from a strong, positive correlation of 0.6 for medical applications to weak correlations of 0.05 in areas of food applications and animal cloning), they also concluded that a "knowledge gap" failed to completely explain the much higher European opposition to biotechnology. In this case, "trust gaps" (i.e., the size of the difference in trust in different stakeholders involved in the technology) emerged as a more reliable predictor of biotechnology attitudes than knowledge levels.

The importance of trust in explaining attitudes, relative to science knowl-

---

[2]The small amounts of variation that were present were explained by the proportion of the population in the countries that went on to attend higher education (Allum et al., 2008).

edge levels, has been stressed elsewhere. Priest (2001), using a path analysis, explored the competing roles of awareness, food safety concerns, genetics knowledge, and trust in key scientific institutions on encouragement for biotechnology, including the genetic engineering of crops, cloning, and engineering bacteria to produce pharmaceuticals, among others. This study revealed a moderate positive relationship between specific knowledge of genetics and encouragement of biotechnology applications. However, the strength of the knowledge-attitude link was much less pronounced than the one found between institutional trust and biotechnology encouragement. Brossard and Nisbet (2007) also found a small but positive relationship between factual knowledge of agricultural biotechnology and support for the technology after controlling for a large number of variables, including sociodemographic variables, media use, levels of trust, and reservations about the effects of science. However, they found that the main determinant of public support for agricultural biotechnology was the level of deference toward scientific authority and not knowledge levels or trust in information providers.

Other scholars have observed negative relationships between various measures of scientific knowledge and public attitudes, particularly for issues characterized by ethical debates (Knight, 2009). For example, Cacciatore and colleagues (2012a) found that, even after controlling for a host of factors, increased knowledge about biofuels was associated with a greater tendency to perceive increased risks relative to benefits from the alternative fuel. Similar patterns have been noted for the issue of nanotechnology, where Lee and colleagues (2005) found that general science knowledge negatively predicted people's perceptions of benefits of the science relative to risks. However, as discussed below, the researchers stressed that processes were complicated and that knowledge had a weaker effect on attitudes for people who showed strong emotional reactions to the topic. In another study, Kahan and colleagues (2012) explored the effects of science literacy and numeracy on climate change attitudes and found that both scientific knowledge and numeracy were associated with decreased risk perceptions regarding the dangers of climate change.

Mediators and Moderators

As illustrated above, there is increasing evidence that the direct link between science knowledge and attitudes toward scientific issues is weak and is mediated or moderated by other factors.[3] Acknowledging that the psychology of attitudes is complex and that cognitive and affective factors have to be taken into account, scholars have explored what factors might shape the connections between knowledge and attitudes. Much of this recent work shows that the

[3]A moderator variable is one that influences the strength of a relationship between two other variables, and a mediator variable is one that explains the relationship between the two other variables.

relationship between knowledge and attitudes often weakens or disappears as additional variables are controlled for in the analysis (e.g., Cacciatore et al., 2011; Hart and Nisbet, 2012; Ho et al., 2008, 2010, 2011; Scheufele et al., 2009). In other words, individuals may make judgments on specific applications of science not based on their knowledge levels, but based on such other factors as people's values (political ideology, religiosity, or deference to science), their level of trust in information providers, or other important variables.

Lee and colleagues (2005) found a moderating effect of negative emotion on the relationship between nanotechnology knowledge and perceptions of the risks relative to benefits of nanotechnology. Specifically, nanotechnology knowledge had a significantly stronger effect on perceptions of risks versus benefits among individuals who reported low levels of negative emotion toward the issue. People without strong negative emotions toward nanotechnology were much less concerned about the risks of nanotechnology as their knowledge levels increased, while those with strong negative emotions were relatively unmoved in their perceptions of risks regardless of their knowledge level. This pattern was also found when general support for nanotechnology was the dependent variable of interest.

Ho and colleagues (2008) noted similar patterns in their work investigating public attitudes toward stem cell research. As with the results for nanotechnology, the authors found that the positive effects of knowledge on stem cell support did not persist once a host of demographic and media use variables were controlled for in the regression model. Consistent with the results noted above, they found that the influence of knowledge on support for embryonic stem cell research was significantly stronger for people low in religiosity in comparison with people high in religiosity. They also found that knowledge had a much stronger relationship with support for stem cell research among liberals than conservatives. Finally, a similar pattern was observed for deference to scientific authority, with knowledge having the strongest effect on support among those reporting high levels of scientific deference.

Kahan and colleagues (2012) investigated two competing hypotheses—what they call the science comprehension hypothesis (that increases in scientific knowledge will lead to greater scientific support) and the cultural cognition theory (that people form their perceptions of risks based on the risk perceptions of those groups with whom they identify)—to explain public attitudes toward climate change. Although they found a negative effect of science literacy and numeracy on climate change concern, they also found that general science knowledge interacted with worldviews in predicting such attitudes. Specifically, knowledge served to polarize the viewpoints of egalitarian communitarians and hierarchical individualists, with increased literacy elevating concern about climate change for the communitarians and decreasing the concern of the individualists.[4] Guy and colleagues (2014) found that knowledge specific to the issue of climate change was associated with an increased tendency to accept the evidence for climate change (a pro-science attitude) among those with a hierarchical worldview, but not among those with an individualistic worldview. These approaches are consistent with "motivated reasoning," the idea that individuals tend to select information that is consistent with their views or beliefs and, alternatively, avoid information that is inconsistent with their views or beliefs (see, e.g., Yeo et al., 2015).

As the work discussed above suggests, the path from scientific knowledge to positive attitudes toward science or support for science is not always clear. Knowledge affects different subgroups in a population differently depending on a host of factors, including levels of religiosity, political predispositions and worldviews, and deference to scientific authority. These patterns seem to vary depending on the specific scientific issue being explored and the culture in which the data is collected. More research is needed to understand this phenomenon.

Effects of Interventions to Increase Knowledge on Attitudes

Experiments related to science knowledge typically seek to assess the effect of providing individuals with new information and comparing the views of those individuals to groups who received either no new information, different information, or some other intervention. However, such experiments can be challenging to interpret. Although effects may emerge, simply learning new facts on their own may not be an adequate representation of the effects of science knowledge or literacy. Specific circumstances may place more or less literacy demands on individuals or allow or disallow for opportunities to apply that literacy. Given this, evidence from interventions that do not take into account context may be limited in its general applicability.

Consistent with similar research on nonscience topics, studies on science-related deliberation clearly show that it is possible to increase basic knowledge through various short-term interventions (Sturgis et al., 2010; Doble, 1995; Einsiedel and Eastlick, 2000; Setälä et al., 2010; Delli Carpini et al., 2004; Bauer and Bonfadelli, 2002). However, the studies also typically show that such learning often has little relationship to attitude change. Gastil and Dillard (1999) found these types of outcomes along with differences by ideology for a range of issues, including energy and health topics. More recently, Kronberger and

[4]Kahan and colleagues (2012) define hierarchical individualists as individuals who tie authority to conspicuous social rankings and eschew collective "interference" with the decisions of individuals possessing such authority. Egalitarian communitarians are defined as individuals who favor less regimented forms of social organization and greater collective attention to individual needs (Kahan et al., 2012, p. 732).

colleagues (2012) found that a reading task followed by focus-group discussions around synthetic biology resulted in increased "opinion certainty," especially for groups whose members were highly interested in the topic.

Science information-related experiments not involving discussion similarly found that providing participants with information has a limited effect or an effect that is contingent on predispositions, such as ideology or worldviews. Druckman and Bolsen (2011) and Bolsen and colleagues (2014), for example, showed that "framing" various technologies in certain ways can equal or overwhelm the effect of providing someone with basic information related to emerging technologies. Furthermore, these studies show that people tend to interpret any new information in a way that fits with their worldviews (i.e., they engage in "motivated reasoning") (see, e.g., Ahern et al., 2016). Kahan and colleagues (2009) have further shown that the process of giving people additional information actually helps people figure out how to use their worldviews to judge a new technology. For example, one study found that views about nanotechnology were initially relatively similar across cultural worldviews prior to receiving risk and benefit information but, once such information was provided, people reacted divergently, in a manner consistent with their different cultural predispositions toward technological risk generally (Kahan et al., 2009, p. 88). Similar effects for issues such as climate change have also been found (Braman et al., 2012).

Brossard and colleagues (2005) used a citizen science project (scientific research conducted, in whole or in part, by amateur or nonprofessional scientists) to explore methods of improving understanding and knowledge of the process of science and bird biology. The study found a significant increase in knowledge of bird biology among those in the treatment condition following the completion of the project. However, corresponding increases in participants' understanding of the scientific process did not occur, and the project did not affect participants' attitudes toward science or the environment.

To summarize, the available evidence suggests that providing people with an opportunity to learn about a topic may result in some learning, but it is unlikely to substantially affect attitudes on scientific issues. The reason that individuals likely do not change their attitudes in response to new information is similar to the reason that variables such as ideology or worldview moderate the relationship between existing science knowledge and attitudes about science. Individuals use both existing and new information to reinforce existing attitudes rather than to change their attitudes. It is also possible that people typically use new information to figure out what their cultural group likely believes. Once they have made this determination, the tendency is to conform to their group's beliefs (Gunther and Liebhart, 2006; Mercier and Sperber, 2011; Hart and Nisbet, 2012; Kahan, 2015).

Because of these phenomena, determining causality—knowing whether knowledge is driving attitudes, whether attitudes are driving people to become more informed, or whether the relationship is reciprocal—can be difficult. Importantly, experimental work, which can shed some light on this issue, has limits; it can be difficult to increase knowledge levels without framing the information in such a way that makes it difficult to pinpoint knowledge, and not the introduction of a specific viewpoint, as the cause of a shift in attitude. In addition, the nature of experiments is such that analyses of knowledge acquisition and subsequent long-term effects on attitudes are difficult to make.

THE RELATIONSHIP BETWEEN SCIENCE LITERACY, HEALTH LITERACY, AND BEHAVIORS

In this section we turn from the effect of knowledge on attitudes to the effects of science knowledge and health knowledge on actions and behaviors. As discussed throughout this report, science literacy and health literacy can operate at many different levels of society, from the actions and decisions of individuals to the collective actions and decisions of a community or even a society. The benefits of science literacy and health literacy can also accrue at each level of society: the boundaries between individual, community, and societal benefits are fluid, and actions or behaviors at different levels may contribute to effects at other levels. For example, the decision to vaccinate one's children is made at the individual level, but it has implications at the level of communities and societies, such as increased life expectancies following the eradication or reduction of infectious diseases.

This section discusses the current evidence on the application of science and health knowledge. We first present frameworks for understanding the relationship between knowledge and action and then analyze the evidence on the relationship between science knowledge, health knowledge, and health-related behaviors. Most of the evidence as to the application of science literacy and health literacy focuses on health-related behavior and does not include a wider set of behavioral outcomes. In reviewing the existing evidence, the committee recognizes that there may be individuals (or communities or societies) who are deeply knowledgeable and engaged, yet nevertheless do not act or take actions that may be at variance with the consensus view of scientists on scientific issues (e.g., individuals objecting to vaccinations or individuals who do not pursue preventative care because they lack access to health care).

Framework of Science Literacy and Health Literacy and Action

Identifying the effects of science literacy or health literacy can be challenging. Often, their effects on any sort of action, decision, or behavior are imagined to be linear, unidirectional, and deterministic: science literacy or health literacy causes desirable outcome X. However, science literacy or health literacy alone is rarely entirely necessary or wholly sufficient for producing a particular desirable outcome. Although a connection between knowledge and actions exists, other factors influence choices—factors ranging from cultural norms and self-efficacy to an individual's ability to access services. Therefore, science literacy and health literacy should be seen as only a probabilistic and partial influence on actions, decisions, and behaviors.[5]

In addition, claims that construe science literacy and health literacy as necessary for particular actions fail to acknowledge the plurality of human motives and life circumstances that can lead to the same outcome. For example, if a person makes healthy dietary choices, it does not necessarily mean that she or he did so because of new nutritional knowledge. Personal habits, social norms, and cultural affiliation can all play a role in shaping behaviors, as can a wide range of beliefs that may or may not reflect or derive from scientific knowledge. The assumption that only scientific knowledge and understanding underlies a particular conclusion, action, or behavior is contradicted by the evidence.[6] Social factors (such as norms, expectations, and regulations) can also shape the resources available to individuals, thus limiting or constraining behavior. For example, in health care systems, the complexity of medical texts, the communication skills of those providing information, and the attributes of institutions that support or impede patients and caregivers and health professionals all shape the behaviors of actors engaged in the system (Pleasant et al., 2016). Furthermore, living in areas with a shortage of health professionals, lacking of health insurance, and facing problems accessing health services may influence health-related behaviors (see e.g., Gore et al., 1999).

Scholars in the social sciences have long worked to conceptualize this problem of human behavior. In general, social scientists argue that behavior may be determined by a range of factors that include: knowledge and skills, perceived risk, attitudes and beliefs, perceived consequences, self-efficacy, social norms, intentions, and demographics. Other social-psychological determinants (e.g., self-concept and self-esteem, occupational stress, religiosity, recreation and leisure, social support networks, and media habits) may also influence a person's actions (see e.g., Ajzen, 1985; Glanz and Rimer, 1995; Prochaska et al., 1992; Bandura, 1971).

The field of health literacy has developed useful frameworks for under-

[5]It is always possible to formulate post hoc arguments about why science literacy and health literacy could not operate in a particular circumstance—even if they were present. It is not the committee's intent to make claims about science literacy and health literacy unfalsifiable. Rather, the committee believes science literacy and health literacy require an assessment of their value on the basis of a range and type of circumstances in which they do appear to shape thoughts, actions, and behaviors in a positive way.

[6]See, for example, Kempton et al. (1995), who found the average knowledge about environmental issues to be low. However, the lack of knowledge was equally strong among environmentalists and nonenvironmentalists, implying that environmental knowledge per se is not a prerequisite for pro- or anti-environmental behavior.

standing individuals' health behaviors and the relationship between a person's health knowledge and their health-related behaviors. In general, these frameworks incorporate internal and external factors that contribute to behavior. For example, variables such as knowledge of opportunities for screening and treatment or an individual's risk perception may spur intentions about a health action and ultimately result in a decision about whether to attempt to perform that health action. Concepts, such as self-efficacy, and practical barriers, like the financial costs associated with the action, influence the translation of intentions into action. Social factors may also influence motivational and volitional processes. In addition to these social cognitive processes, demographic determinants (e.g., gender, employment status, and personal wealth) can influence the likelihood of a health action (see e.g., Paasche-Orlow and Wolf, 2007; von Wagner et al., 2009).

These behavioral theories and frameworks provide a useful guide for thinking about the relationship between science literacy, health literacy, and behaviors or actions. The constraints and, more broadly, the social structures in which individuals live limit their ability to take action on the basis of science literacy or health literacy. Science literacy or health literacy can be a powerful tool, but it can more easily be used by some individuals and in certain circumstances than others. That is, one's science literacy or health literacy may be adequate in certain situations and may be deficient in others. Understanding this relationship allows for a more realistic understanding of the benefits of science literacy and health literacy, even when, at times, that benefit may be limited by other factors.

Science Literacy, Health Literacy, and Behaviors

As discussed above, the connection between an individual's science and health knowledge (as assessed by an individual's knowledge of specific facts) and action is limited. There is particularly strong evidence that knowledge has a mediated relationship to action (e.g., Bord et al., 2000; Kollmuss and Agyeman, 2002).

Science Literacy and Behaviors

Though most of the evidence on use of health literacy examines effects on behaviors related to health, literature in the field of environmental science explores the relationship between knowledge and a nonhealth science-related behavior. In the environmental domain, the available evidence suggests that there is a weak correlation between science knowledge and behavior. For example, Hines and colleagues (1986) published an important meta-analysis that examined 128 pro-environmental behavior research studies. The analysis found that the correlations between knowledge and attitudes, attitudes and intentions, and intentions and actual behavior were weak.

Finger (1994) showed that environmental information and knowledge predicted little of the variability in most forms of environmental behavior. However, the study did find that information and knowledge acquisition appeared to foster protest actions. Hsu (2004) assessed the effects of an environmental education course on students' environmental behavior. The students' perceived knowledge of environmental issues and intentions to act did increase after completing the course, but this study examined only perceived knowledge and intentions to act, not completed behaviors.

There is evidence to suggest that understanding the institutional nature of knowledge production (e.g., peer review, conflicts of interest, research funding) can be useful across a wide range of circumstances (see, e.g., Ryder, 2001). This aspect of science literacy may enable individuals to engage critically with science by helping them to frame meaningful questions, even in contexts in which relevant science concepts are highly technical. Critical understanding of the nature of scientific evidence, a grasp of the way that wider issues influence debates about science, and the value of formal scientific evidence may enable individuals to productively engage with scientific controversies (Tytler et al., 2001). In doing so, individuals may create greater accountability—such as demanding caution in environmental assessments or calling for bioethical guidelines in relation to new genetic technologies—in the production and use of scientific knowledge (Jasanoff, 2003; Irwin and Wynne, 1996).

There are numerous factors beyond scientific knowledge that influence behaviors. Kollmuss and Agyeman (2002) found an individual's ability to undertake environmentally conscious actions was influenced by demographic factors, external factors (e.g., institutional, economic, social, and cultural factors) and internal factors (e.g., motivation, values, attitudes, emotion, locus of control, responsibilities, and priorities). Differences in willingness to act are also mediated by trust and acceptance (Rabinovich et al., 2012). When individuals lack trust in existing political leaders or institutions to respond effectively to climate change, for example, their personal motivation to engage with the issue is dampened, since feelings of reciprocal sacrifice are important motivators of participation (Feldman and Hart, 2016).

Health Literacy and Behaviors Related to Health

The relationship between health literacy, behaviors related to health, and ultimately health outcomes is complex. Much of the research on the relationship between health literacy and behaviors related to health at the individual level focuses on compliance behaviors and the use of health care services. In the field of health literacy, this research tends to use cross-sectional surveys to assess health literacy of a sample of respondents and measure various outcomes. It is common for surveys to use "fundamental" literacy as a proxy for health literacy or instruments such as the Test of Functional Health Literacy in Adults, the Rapid Estimate of Adult Literacy in Medicine, or the Wide Range Achievement Test to assess an individual's health literacy (see Chapter 2). The outcomes that are assessed range from compliance behaviors, such as adherence to oral contraceptive pills or rate of breastfeeding, to self-efficacy and health outcomes, such as healthy weight, diabetes control, or management of asthma (Ross et al., 2001; DeWalt et al., 2007; Hawthorne, 1997; Davis et al., 2006; Campbell et al., 2004; DeWalt and Hink, 2009).

It is well accepted that health literacy is an important factor in patients' abilities to obtain and use health-related information to make decisions about health care as it relates to health services utilization, self-care behaviors, and risks for disease-related morbidities and mortalities, especially when tied to racial and ethnic disparities (DeWalt et al., 2004; Berkman et al., 2011a, 2011b; Osborn et al., 2007, 2011; Paasche-Orlow et al., 2010). For example, in a systematic review the Agency for Healthcare Research and Quality (2011) found evidence that lower levels of health literacy were associated with increased hospitalization, greater emergency care use, lower use of mammography, and lower receipt of the influenza vaccine (see also Berkman et al., 2011a, 2011b; Sheridan et al., 2011).

Although much of this research demonstrates a relationship between knowledge and action, it is typically weak (e.g., Al Sayah et al., 2012). Studies often find that other factors, such as reasoning skills, intelligence, trust, and values, contribute to an individual's actions (see, e.g., Arnold et al., 2001; Keller et al., 2008). Other mediating factors may be external to the individual, such as access to health care, patient-provider interactions, cost, and management of health and illness (von Wagner et al., 2009). For example, in a meta-analysis reviewing the relationship between health literacy and medication adherence, Zhang and colleagues (2014) found higher health literacy levels were associated with better medication adherence across 6 diseases and 35 samples. However, the effect size of the relationship was weak when compared with other predictors of medication adherence, such as type of disease, medication beliefs, and cost restraints (Zhang et al., 2014).

Some studies have examined interventions intended to improve health behaviors and use of health services. For example, Yin and colleagues (2008) measured parent-reported medication dosing and observed parents preparing a medication dose. Parents in the intervention group received pictogram-based medication instruction sheets with teach-back counseling; parents in the control group received standard care. The parents in the intervention group were more likely to use the correct dose and had greater self-reported adherence to the prescribed medication regimen. In another study, Robinson and colleagues (2008) studied children with asthma who were enrolled in a reading skills and asthma education program. Of the children enrolled, 63 percent visited an emergency room before the intervention while only 33 percent made an emergency visit during the intervention, and the children whose reading improved the most were least likely to have repeated emergency visits. In contrast, a recent study of parents who received one of four interventions—information explaining the lack of evidence linking vaccinations to autism; textual information about the dangers of diseases prevented by vaccination; images of children who have diseases prevented by the vaccine; and a dramatic narrative about an infant who almost died of a disease prevented by the vaccine—found that none of the interventions increased parental intent to vaccinate a future child. In addition, the results indicated that showing images of sick children increased expressed belief in a vaccine-autism link and the dramatic narrative increased self-reported belief in serious vaccine side effects (Nyhan et al., 2014).

The field of health literacy has acknowledged that many factors—at both the individual level and the system level—may affect an individual's application of health literacy; see Box 5-3 for a discussion of the relationship between health literacy and health disparities. System factors shape the resources available to and the behaviors of visitors and users, as well as the behaviors and expectations of the broad array of individuals engaged in providing health information, care, and services (Pleasant et al., 2016, p. 3). Studies of interventions demonstrate that, at least in the short term, an individual's health literacy may be enhanced by an outside intervention, typically a change in the health system. These interventions contribute to literacy and may spur particular behaviors if they can overcome adverse individual and external influences and potential context-based hurdles.

The Effects of Science Literacy on Behaviors Related to Health

As described above, the relationship between health literacy, behaviors related to health, and ultimately health outcomes is complex. Although the effect of health literacy on health outcomes is widely accepted, it is unlikely that health literacy, any more than science literacy, has direct effects on most health outcomes, as this relationship is mediated by a range of factors.

Beyond this literature, there has been little scholarly work analyzing the relationship between science literacy and health-related behaviors. However, studies do show that science knowledge can lead to increased perceived self-efficacy, and it is well established that perceived self-efficacy and response efficacy are important in explaining preventive health care behaviors (e.g., Jayanti and Burns, 1998; Bandura, 2010). Thus, in this way, science literacy may indirectly affect preventive health behaviors. Despite this finding, the available evidence does not demonstrate that science literacy is directly related to specific health behaviors. If enhanced science literacy can be empirically linked to enhanced health literacy, it could indirectly be associated with some behaviors related to health. There is a need for further research to determine if such a relationship exists.

One such promising area is the potential relationship between science

BOX 5-3
Will Improving Health Literacy Reduce Health Disparities for Vulnerable Populations?

The text below is from a presentation by Dean Schillinger, summarized in Institute of Medicine (2011, pp. 16-17).

Turning to the question of whether health literacy intervention reduces disparities, there is much less data. Most studies evaluating health literacy interventions have demonstrated improvements that disproportionately accrue to those with adequate health literacy, or they yield similar improvements across health literacy. Most studies do not report the effects on vulnerable subgroups by, for example, stratifying results by race, ethnicity, or educational attainment.

A few health literacy interventions have been found to disproportionately affect vulnerable subgroups. Rothman and colleagues (2004) tested a health-literacy-sensitive diabetes management program that disproportionately benefited those with limited health literacy compared with those who had adequate health literacy. DeWalt and colleagues (2006) did the same on congestive heart failure. Paasche-Orlow and colleagues (2005) used a teach-to-goal approach in asthma education, which disproportionately benefited those with limited health literacy versus those with adequate health literacy.

An automated diabetes phone system disproportionately engaged and led to behavior change among those with limited literacy and limited English proficiency compared with others (Schillinger et al., 2008). Finally, work by Machtinger and colleagues (2007) found visual medication schedules, when combined with a "teach back" in anticoagulation care, disproportionately benefited those with communication barriers.

Reducing disparities requires taking a socioecological approach, Schillinger said. This approach includes thinking about the context in which people live and receive their healthcare. An important question to consider is, will better individual health literacy lead people to make healthier choices, particularly those who are in vulnerable populations? Given the cluster of risks that vulnerable populations face—food insecurity, food access problems, unsafe neighborhoods, and so on—that are determined by social context, it is important to be realistic regarding expectations of what improving health literacy can do.

Another important issue is whether attempts to affect individual health literacy will be hampered by the nature of health systems that disproportionately care for vulnerable populations. Varkey and colleagues (2009) studied primary care practices stratified by the proportion of minority patients served. Practices that served 30 percent or more minority patients were compared to practices with fewer minorities. Tremendous differences were found in organizational structure, workforce satisfaction, comorbidity, complexity of disease, and perceived practice chaos. Work settings significantly affect the health care provided to vulnerable populations.

Context is important. Preliminary evidence suggests these factors can affect the relationship between health literacy and outcomes. In studies of the relationship between health literacy and chronic disease control (e.g., blood pressure, diabetes), whether or not there is a relationship between health literacy and the outcome appears to vary based on the setting. Schillinger's work at a public hospital showed a clear relationship between health literacy and diabetes outcomes, but a similar replication study in a private setting in New England found no relationship (Morris et al., 2006) and a study of literacy and blood pressure control found that the relationship varied by setting (Powers et al., 2008). Contextual factors need further analysis.

literacy and information-seeking and interpretation of online health-related information. Individuals now routinely seek health-related information online. When doing so, individuals need to possess the ability to determine which information is trustworthy and process the appropriate information in an efficient manner. There is some evidence to suggest that science literate individuals may be more equipped to efficiently look for and process health-related information (Ellis et al., 2012).

CONCLUSIONS

The existing empirical evidence at the individual level on the value of science literacy is drawn largely from two separate research fields: science literacy and health literacy. Studies on the effect of health literacy have largely examined the relationship between knowledge (measured as health-related knowledge or foundational literacy) and behaviors related to health. In contrast, most of the literature on science literacy assesses the relationship between science knowledge and attitudes toward, perceptions of, and support for science.

> **CONCLUSION 10** Research examining the application of science literacy and health literacy has focused on different things: studies on the impact of health literacy have looked for impact on health-related behaviors and actions (e.g., compliance with medical advice, shared decision making, etc.), whereas studies on the impact of science literacy have mostly examined its relationship to individual attitudes toward science and support for scientific research.

Findings from regression-based analyses—in which the effects of knowledge on attitudes were often weakened or eliminated completely when demographic, value predisposition, media use, and trust variables were included in the analysis—demonstrate that context matters when looking at the relationship between knowledge (as assessed by currently used measures of scientific content knowledge) and perceptions of and support for science. Though there appears to be a small, positive relationship between general science knowledge and general attitudes toward science, the relationship is weak and is moderated or mediated by other factors. Therefore, given the state of current evidence and measures, increasing science literacy should not be seen as the foremost means for improving support for science.

> **CONCLUSION 11** Available research does not support the claim that increasing science literacy will lead to appreciably greater support for science in general.

Scholars have shown that the relationship between knowledge and attitudes becomes complicated when assessing science knowledge and attitudes toward specific science issues. Knowledge affects diverse subgroups in the population differently depending on a host of factors, including levels of religiosity, political predispositions and worldviews, and deference to scientific authority. These patterns seem to vary depending on the specific scientific issue being explored and the culture in which the data are collected. In fact, there is often an interaction between knowledge and worldviews such that enhanced knowledge has been associated, in cases of controversial issues, with increased polarization, affecting attitudes toward those specific science issues.

> **CONCLUSION 12** Measures of science literacy in adult populations have focused on a very limited set of content and procedural knowledge questions that have been asked within the constraints of large population surveys. Though available measures are limited in scope, evidence suggests they are reasonable indicators of one aspect of science literacy, science knowledge. Studies using these measures observe a small, positive relationship between science literacy and attitudes toward and support for science in general.
>
> > **CONCLUSION 12a** An individual's general attitude toward science does not always predict that same individual's attitude toward a specific science topic, such as genetic engineering or vaccines.
> >
> > **CONCLUSION 12b** Some specific science issues evoke reactions based on worldviews (e.g., ideology, religion, deference to scientific authority) rather than on knowledge of the science alone.

The research on the relationship between science literacy, health literacy, and behaviors related to health is limited, but the examples that exist highlight the weak correlation between science literacy and health literacy and behaviors. Like the relationship between science knowledge and attitudes toward science, the causal pathway between science literacy and health literacy and behaviors is complex and mediated or moderated by personal and external factors.

> **CONCLUSION 13** The commonly used measures of science and health literacy, along with other measures of scientific knowledge, are only weakly correlated with action and behavior across a variety of contexts.

The weak relationships among science literacy, health literacy, attitudes, and behaviors suggest that efforts to simply promote knowledge and understanding to change behavior or attitudes may have limited results. Efforts that focus on increasing knowledge also need to include removing impediments to actions and lowering the literacy demands of particular situations.

6

Research Agenda

Throughout this report, the committee has both illuminated and complicated evolving notions of science literacy. This analysis reflects our understanding of the history of how the term has been understood and applied, as well as suggests a new direction for future thinking on these issues. In order to continue the trajectory of this work, the committee offers a recommendation for how the term should be considered and applied moving forward.

> **Recommendation:** The committee recommends that, in keeping with contemporary thinking, the scientific community, the research community, and other interested stakeholders continue to expand conceptions of science literacy to encompass (a) an understanding of how social structures might support or constrain an individual's science literacy and (b) an understanding that societies and communities can demonstrate science literacy in ways that go beyond aggregating the science literacy of the individuals within them.

In continuing to expand the conceptions of science literacy, it will be necessary to solidify an evidence base that investigates science literacy in all its complexity.

> **Recommendation:** The committee recommends that the research community take on a research agenda that pursues new lines of inquiry around expanding conceptions of science literacy.

This chapter outlines some areas of research needs. We offer sets of research questions (found throughout this chapter in bold) as a way of thinking about creating new measures and expanding the information available. These questions, divided into four issues, reflect the need to better understand (1) the relationship between science knowledge and attitudes toward science, (2) the utility of science literacy, (3) the relationship of science literacy to other literacy skills, and (4) the role of science literacy for citizens as decision makers.

ISSUE 1: KNOWLEDGE AND ATTITUDES

In our review of the literature on science literacy, we conclude that the research shows a positive relationship between science knowledge and attitudes toward science and support for science. This finding stems largely from decades of survey research assessing the public knowledge and general attitudes toward science. This relationship, however, is small and mediated such that analyses of the survey data suggest that an increase in knowledge across a population is unlikely to result in an appreciable change in positive attitude toward science.

Interventions intended to change individuals' attitudes often have focused on providing scientific information to increase individuals' knowledge. Experimental and case studies of such interventions have sought to assess the effect of providing individuals with new information, comparing the views of individuals who were given specific information to individuals who received either no new information, different information, or some other intervention. These experiments can be challenging to interpret. While some limited effects have been observed, it is not clear that short-term learning of new facts is the same as science literacy or the development of meaningful knowledge. In addition, evidence from interventions that do not take into account context may have limited applicability in other contexts. More research is needed to understand the impact of efforts to enhance science knowledge and their effects on attitudes and behaviors, as well as the role of context and the relationship between findings from experimental and case studies and those from cross-sectional surveys.

1.1 Under what conditions and for which types of knowledge does acquiring new scientific knowledge affect individual attitudes and behavior related to science? In other words, to what degree are the results of experimental interventions consistent with results that emerge from studies based on cross-sectional surveys?

The potential relationship between science literacy and information-seeking/interpretation behaviors, such as searching online information or participating in research, needs to be investigated. Individuals now routinely seek science-related and health-related information online. When doing so, individuals need to possess the ability to determine which information is trustworthy and process the appropriate information in an efficient manner. There is some evidence to suggest that science literate individuals may be more equipped to efficiently look for and process health-related information (see Chapter 5). Life circumstances including (but not limited to) group membership and geographic location affect what resources people can access and use. Despite a remarkable leveling in access to the Internet and Internet-enabled technologies in the past 2 decades, research continues to point to troubling differences in the way that people are supported in their use of Internet technologies, particularly in educational settings (see Chapters 3 and 5).

1.2 How do developments in digital resources and growth in participatory research opportunities (e.g., citizen science) change information seeking in, understanding of, and attitudes toward science? Particular attention should be placed on understanding access and use of information (i.e., factors that play into access to information, information-seeking habits, and evaluation and use of information.)

ISSUE 2: UTILITY OF SCIENCE LITERACY

Currently science literacy is assessed in adult populations throughout the world using survey instruments that reflect knowledge of content and some limited ways of scientific thinking. These instruments are widely used and methodologically sound, but constraints on length and demands for comparability across nations mean that they may not capture as much of the depth and breadth of the knowledge nor the full diversity of scientific reasoning required for science literacy. Researchers in the field need to come to a common understanding of potential indicators as well as limitations of what can be measured regarding the utility of science literacy. More research is needed to produce new measures to better examine how science literacy at society, community, and individual levels is shaped by and contributes to behaviors, attitudes, and cultures.

2.1 To what extent do the current measurements of science literacy map onto people's capacities to accomplish specific tasks, such as to understand science or health messages, choose between competing sources of information, identify expertise, or modify behavior? For example, what can someone who scores in the upper quartile on a science literacy measure do that someone who scores in the lowest quartile cannot? How do these scores relate to the probability of success in either science, technology, engineering, and mathematics (STEM) or STEM-enabled jobs? How does science knowledge—and knowledge about science—shape a person's ability and willingness to engage with contemporary scientific and technological issues?

Case studies have demonstrated instances in which communities became centrally involved in the interpretation of scientific research or critically engaged in community-based decision making on science-related issues. However, the current literature may be biased toward successful examples and thus miss instances in which science literacy did not emerge or in which a community was not able to develop the capacity to manage a scientific issue it faced. In addition, the role of science literacy in accomplishing community goals has typically been extrapolated from findings afterward and not considered at the outset of studies. Future research should consider explicitly the nature of science literacy within communities. In most cases there has been little effort to assess, prior to the start of community action or controversy, what level of science literacy was present in the community. Though, we recognize that many of the concepts that drive community-level analyses of science literacy may be difficult to operationalize empirically.

2.2 How can research measure, understand, and support the features, structures, and circumstances of communities that make it possible for them to engage collectively with and use science? Research on this topic requires comparisons across multiple communities.

The context and demands for science literacy at every level of social organization—society, community, and individual—are variable and may shift as new scientific advances and discoveries emerge. Empirical findings suggest that the knowledge needed to engage with science in contemporary societies is somewhat different than the specific content knowledge captured by existing measures of science literacy. Similarly, researchers have observed that existing measures of health literacy do not address the full conceptualization of health literacy as has developed in the field in recent years (see Chapter 2).

The growth in communications media is one example of the complexity of information sharing that could affect science literacy and of an area that deserves more research attention. The breadth of the Internet's topical coverage and the increasing ease of access through mobile devices have changed how people think about and use information and interact with others. To date, what little data exists from these media suggest that people participate in large numbers and that participation spans national boundaries. Recent figures estimate that 84 percent of U.S. citizens use the Internet, but participation is not uniform. Nearly all young adults, those with higher levels of education, and those in the most affluent households use the Internet. Use among other age, education, and income subgroups is rising but gaps in use remain.[1] The Internet and related media make possible the formation of virtual communities. Evolving communication technologies allow people from all over the world to

[1]See, for example, Perrin and Duggan (2015).

exchange information and to connect with others on issues of interest and/or relevant to them. These virtual communities could have significant as well as disparate impact across societies on individuals' information seeking and informal learning.

There are currently few data from these media sources that bear on science literacy and health literacy. However, there is opportunity to create datasets and examine the collective action made possible by virtual environments and social media and investigate whether beneficial community-level or national-level knowledge is made possible by them. As more information becomes available, researchers could examine the contribution to science literacy at multiple levels and how such literacy becomes distributed.

2.3 To what degree are existing measures of science and health literacy associated with knowledge about emerging issues? To answer this question, the research community should continue to develop, test, and validate measurement tools that assess science literacy and health literacy.

2.4 Given the complex, variable, and contingent nature of the situations in which people develop and use science literacy, what new research tools are needed to complement commonly used survey-based measures?

ISSUE 3: OTHER LITERACY SKILLS

Making progress toward understanding the constraints on and supports for achieving adequate levels of science literacy requires understanding better the relationships among science literacy, health literacy, and foundational literacy. Foundational literacy encompasses the skills and capacities necessary to process and be fluent in the use of words, language, numbers, and mathematics. The committee recognizes that all other domains of literacy depend on foundational literacy. New forms of domain literacy emerge when an individual or group attempts to identify a particular set of knowledge or competencies as socially important. Science literacy and health literacy have been the subject of concerted scholarly attention, albeit in separate research communities.

Health literacy appears closely related and somewhat overlapping with science literacy since science content areas, such as biology or chemistry, are necessary for understanding basic health concepts. One could envision that some level of science literacy is essential for performing the knowledge, skills, and fluency necessary to be health literate. However, there is relatively little empirical work explaining this relationship or analyzing the relationship between science literacy and health-related behaviors.

3.1 What is the relationship among different types of literacies, including foundational literacy, science literacy, and health literacy? Are there threshold

levels of foundational literacy required for accessing health literacy or science literacy? How do these different types of literacies relate to attitudes and behaviors that are related to health and science?

Although it is rarely discussed in cross-national studies of science literacy, there is often greater variation within a country than between countries. A closer look at variation in the United States reveals stark disparities in knowledge, access to knowledge, and access to systems (e.g., education, health care, and justice) that enable people to interpret and act on the knowledge they have. Research on stratification, variation, and disparities in science literacy is rare and more work is needed to understand the association between differential access and the development and use of science literacy. A thorough account of disparities in science literacy cannot rely on individual measures alone; it must examine the conditions that structure science literacy differently for different groups of people, shaping how their understandings of science are developed and deployed.

3.2 How is science literacy distributed within society, and what broader societal factors affect how people access, develop, and use science literacy? What are the disparities in the distribution of science literacy associated with race, ethnicity, gender, schooling, or geographic region? How might such disparities be mitigated?

3.3 How can research measure science literacy and health literacy in the context of the constraints that the broader social systems place on the individual and communities, and the opportunities that those systems provide?

ISSUE 4: CITIZENS AS DECISION MAKERS

It is generally assumed that improving civic science literacy is a social good, regardless of its effect on support for funding of scientific research. The value of science literacy in societal systems such as the health care system, the justice system, and the various systems of political participation, as well as the opportunities that these systems provide to develop science literacy, have not been studied in sufficient detail. What is known, however, is sufficient to conclude that different individuals and communities in a society need different levels of science literacy at different times to accomplish their personal and civic goals. More research is needed to better understand the role of science literacy for citizens as decision makers and consumers of science.

4.1 Participation in particular social systems requires different, perhaps deeper levels of science literacy. For example, citizens participating in the legal system (judges, lawyers, jurors, plaintiffs, defendants) may require different

understanding of scientific concepts for justice to be served. Research on science literacy should also examine the particular demands of participation in critical social systems. Where the legal system is concerned, it is particularly important to know what fields of science are most frequently referenced in the legal arena and what level of understanding of scientific principles, methodologies, and habits of mind are needed for the proper and equitable operation of the justice system.

4.2 Many people not employed in science, technology, engineering, and mathematics (STEM) or STEM-related occupations nonetheless watch television programs about science, read science magazines and books, frequent natural history museums, participate in citizen science and other community science-related activities, and in other ways appreciate science in the same way people appreciate art, music, and literature. Is a higher level of knowledge about science and its methodology associated with increased appreciation of science? To what degree does taking advantage of opportunities to appreciate science increase science literacy?

NEXT GENERATION OF RESEARCH ON SCIENCE LITERACY

This committee was asked to review existing research literature and metrics on science literacy. Much of the current literature focuses on examining the relationship between science knowledge and attitudes toward science using data from large population surveys measuring individuals' understanding of and factual knowledge and scientific processes. Research on individual-level science literacy provides invaluable insights, but on its own offers an incomplete account of the nature, development, distribution, and impacts of science literacy within and across communities and societies.

The literature posits arguments both for individuals and societies on the value of science literacy. However, the research community has yet to study in sufficient detail the value of science literacy "in action" in society and within societal systems and communities. Furthermore, the committee recognizes that social systems such as the health care and education systems provide opportunities to develop science literacy and that the structures within these systems may enable or constrain the development of science literacy. These ideas should be examined in depth, and investigations should pay particular attention to impacts on different subgroups of the population. With new lines of inquiry and development of a wider range of metrics, the complex nature of science literacy may be better understood and put to use. With different levels of analyses, the research community may discover important interconnections between science literacy in society, in communities, and in individuals.

References

Agency for Healthcare Research and Quality. (2011). *Health Literacy Interventions and Outcomes: An Updated Systematic Review.* Rockville, MD: Agency for Healthcare Research and Quality.

Agrawal, A. (1995). Dismantling the divide between indigenous and scientific knowledge. *Development and Change, 26*(3), 413-439.

Ahern, L., Connolly-Ahern, C., and Hoewe, J. (2016). Worldviews, issue knowledge, and the pollution of a local science information environment. *Science Communication, 38*(2), 228-250.

Ajzen, I. (1985). From intentions to actions: A theory of planned behavior. In J. Kuhl and J. Beckman (Eds.), *Action-Control: From Cognition to Behavior* (pp. 11-39). Heidelberg, Germany: Springer.

Al Sayah, F., Majumdar, S., Williams, B., Robertson, S., and Johnson, J. (2012). Health literacy and health outcomes in diabetes: A systematic review. *Journal of Internal Medicine, 28*(3), 444-452.

Allen, B.L. (2003). *Uneasy Alchemy: Citizens and Experts in Louisiana's Chemical Corridor Disputes.* Cambridge, MA: MIT Press.

Allum, N., Sturgis, P., Tabourazi, D., and Brunton-Smith, I. (2008). Science knowledge and attitudes across cultures: A meta-analysis. *Public Understanding of Science, 17*(1), 35-54.

American Association for the Advancement of Science. (1989). *Science for All Americans: A Project 2061 Report on Literacy Goals in Science, Mathematics and Technology.* Washington, DC: American Association for the Advancement of Science.

American Association for the Advancement of Science. (1993). *Benchmarks for Scientific Literacy.* Washington, DC: American Association for the Advancement of Science.

American Medical Association. (1999). Health literacy: Report of the Council on Scientific Affairs. *Journal of the American Medical Association, 281*(6), 552-557.

Ancker, J.S., and Kaufman, D. (2007). Rethinking health numeracy: A multidisciplinary literature review. *Journal of the American Medical Informatics Association, 14*(6), 713-721.

Anderson, M. (2015). *The Race Gap in Science Knowledge.* Washington, DC: Pew Research Center. Available: http://www.pewresearch.org/fact-tank/2015/09/15/the-race-gap-in-science-knowledge [June 2016].

Arnold, C., Davis, T., Berkel, H., Jackson, R., Nandy, I., and London, S. (2001). Smoking status, reading level, and knowledge of tobacco effects among low-income pregnant women. *Preventive Medicine, 32*(4), 313-320.

Bäckstrand, K. (2003). Civic science for sustainability: Reframing the role of experts, policy makers, and citizens in environmental governance. *Global Environmental Politics, 3*(4), 24-41.

Bairoch, P. (1982). International industrialization levels from 1750 to 1980. *Journal of European Economic History, 11*(2), 269-325.

Baker, D.W., Williams, M.V., Parker, R.M., Gazmararian, J.A., and Nurss, J. (1999). Development of a brief test to measure functional health literacy. *Patient Education and Counseling, 38*(1), 33-42.

Bandura, A. (1971). *Social Learning Theory*. New York: General Learning Press.

Bandura, A. (2010). Self-efficacy. In I.B. Weiner and W.E. Craighead (Eds.) *The Corsini Encyclopedia of Psychology* (4th ed., pp. 1534-1536). Hoboken, NJ: John Wiley & Sons.

Bang, M., and Medin, D. (2010). Cultural processes in science education: Supporting the navigation of multiple epistemologies. *Science Education, 94*(6), 1008-1026.

Barron, B., Gomez, K., Pinkard, N., and Martin, C.K. (2014). *The Digital Youth Network: Cultivating Digital Media Citizenship in Urban Communities*. Cambridge, MA: MIT Press.

Barton, A.C. (2003). *Teaching Science for Social Justice*. New York: Teachers College Press.

Bauer, M.W. (2009). The evolution of public understanding of science—Discourse and comparative evidence. *Science, Technology, & Society, 14*(2), 221-240.

Bauer, M.W., and Bonfadelli, H. (2002). Controversy, mass media coverage, and public knowledge. In M.W. Bauer and G. Gaskell (Eds.), *Biotechnology: The Making of a Global Controversy* (pp. 149-175). Cambridge, UK: Cambridge University Press.

Bauer, M.W., and Falade, B.A. (2014). Public understanding of science: Survey research around the world. In M. Bucchi and B. Trench (Eds.), *Routledge Handbook of Public Communication of Science and Technology* (2nd ed., pp. 140-159). New York: Routledge.

Bauer, M.W., Durant, J., and Evans, G. (1994). European public perceptions of science. *International Journal of Public Opinion Research, 6*(2), 163-186.

Bauer, M.W., Durant, J., and Gaskell, G. (1997). Europe ambivalent on biotechnology. *Nature, 387*, 845-847.

Bauer, M.W., Petkova, K., and Boyadjieva, P. (2000). Public knowledge of and attitudes to science: Alternative measures that may end the "science war." *Science Technology & Human Values, 25*(1), 30-51.

Bauer, M.W., Allum, N., and Miller, S. (2007). What can we learn from 25 years of PUS survey research? Liberating and expanding the agenda. *Public Understanding of Science,* 16(1), 79-95.

Bauer, M.W., Shukla, R., and Kakkar, P. (2012a). *Public Understanding of Science in Europe 1989-2005.* Mannheim, Germany: GESIS. Available: http://www.gesis.org/eurobarometer-data-service/search-data-access/eb-trends-trend-files/eb-pus-1989-2005 [July 2016].

Bauer, M.W., Shukla, R., and Allum, N. (2012b). Towards cultural indicators of science with global validity. In M.W. Bauer, R. Shukla, and N. Allum (Eds.), *The Culture of Science: How the Public Relates to Science Across the Globe* (pp. 39-54). New York: Routledge.

Beaton, A.E., Martin, M.O., Mullis, I.V.S., Gonzalez, E.J., Smith, T.A., and Kelly, D.L. (1996). *Science Achievement in the Middle School Years: IEA's Third International Mathematics and Science Study (TIMSS).* Chestnut Hill, MA: TIMSS International Study Center, Boston College.

Bereiter, C. (2002). Liberal education in a knowledge society. In B. Smith (Ed.), *Liberal Education in a Knowledge Society* (pp. 11-34). Chicago, IL: Open Court.

Berkes, F. (2009). Indigenous ways of knowing and the study of environmental change. *Journal of the Royal Society of New Zealand, 39*(4), 151-156.

Berkman, N.D., Dewalt, D.A., and Pignone, M.P. (2004). Literacy and health outcomes. *Evidence Report/Technology Assessment, 87*, 1-8.

Berkman, N.D., Sheridan, S., and Donahue, K. (2011a). *Health Literacy Interventions and Outcomes: An Updated Systematic Review.* Evidence Report/Technology Assessment No. 199. Rockville, MD: Agency for Healthcare Research and Quality. Available: http://www.ncbi.nlm.nih.gov/books/NBK82434 [June 2016].

Berkman, N.D., Sheridan, S., Donahue, K., Halpern, D., and Crotty, K. (2011b). Low health literacy and health outcomes: An updated systematic review. *Annals of Internal Medicine, 155*(2), 97-107.

Besley, J.C. (2010). Public engagement and the impact of fairness perceptions on decision favorability and acceptance. *Science Communication, 32*(2), 256-280.

Besley, J.C. (2016). The National Science Foundation's science and technology survey and support for science funding, 2006-2014. *Public Understanding of Science*, e1-e16.

Besley, J.C., and Oh, S.-H. (2014). The impact of accident attention, ideology, and environmentalism on American attitudes toward nuclear energy. *Risk Analysis, 35*(5), 949-964.

Bohensky, E., and Maru, Y. (2011). Indigenous knowledge, science, and resilience: What have we learned from a decade of international literature on "integration." *Ecology and Society, 16*(4), 6.

Bolsen, T., Druckman, J., and Cook, F. (2014). How frames can undermine support for scientific adaptations: Politicization and the status-quo bias. *Public Opinion Quarterly, 78*(1), 1-26.

Bonney, R., Cooper, C.B., Dickinson, J., Kelling, S., Phillips, T., Rosenberg, K.V., and Shirk, J. (2009). Citizen science: A developing tool for expanding science knowledge and scientific literacy. *BioScience, 59*(11), 977-984.

Bonney, R., Shirk, J.L., Phillips, T.B., Wiggins, A., Ballard, H.L., Miller-Rushing, A.J., and Parrish, J.K. (2014). Next steps for citizen science. *Science, 343*(6178), 1436-1437.

Bord, R., O'Connor, R., and Fisher, A. (2000). In what sense does the public need to understand global climate change? *Public Understanding of Science, 9*(3), 205-218.

Boston University. (2016). *Health Literacy Tool Shed.* Boston, MA: Boston University. Available: http://healthliteracy.bu.edu [July 2016].

Braman, D., Kahan, D., Jenkins-Smith, H., Tarantola, T., and Silva, C. (2012). Geoengineering and the science communication environment: A cross-cultural experiment. *GW Law Faculty Publications & Other Works,* 199.

Brossard, D. (2013). New media landscapes and the science information consumer. *Proceedings of the National Academy of Sciences, 110*(Suppl. 3), 1-6.

Brossard, D., and Nisbet, M.C. (2007). Deference to scientific authority among a low-information public: Understanding U.S. opinion on agricultural biotechnology. *International Journal of Public Opinion Research, 19*(1), 24-52.

Brossard, D., and Scheufele, D.A. (2013). Science, new media, and the public. *Science, 33*(6115), 40-41.

Brossard, D., and Shanahan, J. (2006). Do they know what they read? Building a scientific literacy measurement instrument based on science media coverage. *Science Communication, 28*(1), 47-63.

Brossard, D., Lewenstein, B., and Bonney, R. (2005). Scientific knowledge and attitude change: The impact of a citizen science project. *International Journal of Science Education, 27*(9), 1099-1121.

Brossard, D., Scheufele, D.A., Kim, E., and Lewenstein, B.V. (2009). Religiosity as a perceptual filter: Examining processes of opinion formation about nanotechnology. *Public Understanding of Science, 18*(5), 546-558.

Brown, P. (1992). Popular epidemiology and toxic waste contamination: Lay and professional ways of knowing. *Journal of Health and Social Behavior, 33*(3), 267-281.

Brown, P. (1993). When the public knows better: Popular epidemiology challenges the system. *Environment: Science and Policy for Sustainable Development, 35*(8), 16-41.

Brulle, R.J., and Pellow, D.N. (2006). Environmental justice: Human health and environmental inequalities. *Annual Review of Public Health, 27*, 103-124.

Bucchi, M., and Neresini, F. (2008). Science and public participation. In E.J. Hackett, O. Amsterdamska, M. Lynch, and J. Wajcman (Eds.), *The Handbook of Science and Technology Studies* (3rd ed., pp. 449-473). Cambridge, MA: MIT Press.

Bybee, R., McCrae, B., and Laurie, R. (2009). PISA 2006: An assessment of scientific literacy. *Journal of Research in Science Teaching, 46*(8), 865-883.

Cacciatore, M.A., Scheufele, D.A., and Corley, E.A. (2011). From enabling technology to applications: The evolution of risk perceptions about nanotechnology. *Public Understanding of Science, 20*(3), 385-404.

Cacciatore, M.A., Binder, A.R., Scheufele, D.A., and Shaw, B.R. (2012a). Public attitudes toward biofuels: Effects of knowledge, political partisanship, and media use. *Politics and the Life Sciences, 31*(1-2), 36-51.

Cacciatore, M.A., Scheufele, D.A., and Shaw, B.R. (2012b). Labeling renewable energies: How the language surrounding biofuels can influence its public acceptance. *Energy Policy, 51*, 673-682.

Campbell, D.T., and Stanley, J.C. (1963). *Experimental and Quasi-Experimental Designs for Research*. Boston, MA: Houghton Mifflin.

Campbell, F., Goldman, B., Boccia, M., and Skinner, M. (2004). The effect of format modifications and reading comprehension on recall of informed consent information by low-income parents: A comparison of print, video, and computer-based presentations. *Patient Education and Counselling, 53*(2), 205-216.

Carl, N., and Cofnas, N. (2016). Scientific literacy, optimism about science and conservatism. *Personality and Individual Differences, 94*, 299-302.

Carnoy, M., and Rothstein, R. (2013). *What Do International Tests Really Show about U.S. Student Performance?* Washington, DC: Economic Policy Institute.

Cashman, S., Adeky, S., Allen, A.J., Corburn, J., Israel, B., Montano, J., Rafelito, A., Rhodes, S., Swanston, S., Wallerstein, N., and Eng, E. (2008). The power and the promise: Working with communities to analyze data, interpret findings, and get to outcomes. *American Journal of Public Health, 98*(8), 1407-1417.

Center on Education Policy. (2007). *Choices, Changes, and Challenges: Curriculum and Instruction in the NCLB Era*. Washington, DC: Center on Education Policy.

Chin, M.H., Walters, A.E., Cook, S.C., and Huang, E.S. (2007). Interventions to reduce racial and ethnic disparities in healthcare. *Medical Care Research and Review, 64*(Suppl. 5), S7-S28.

China Research Institute for Science Popularization. (2010). *Chinese Public Understanding of Science and Attitudes towards Science and Technology*. Beijing: China Research Institute for Science Popularization.

Clapp, R.W. (2002). Popular epidemiology in three contaminated communities. *The ANNALS of the American Academy of Political and Social Science, 584*(1), 35-46.

Coben D. (2000). Numeracy, mathematics and adult learning. In I. Gal (Ed.), *Adult Numeracy Development: Theory, Research, Practice* (pp. 33-50). Cresskill, NJ: Hampton Press.

Coben D. (2003). *Adult Numeracy: Review of Research and Related Literature*. London: National Research and Development Centre for Adult Literacy and Numeracy.

Cohn, J.P. (2008). Citizen science: Can volunteers do real research? *BioScience, 58*(3), 192-197.

Committee on the Objectives of a General Education in a Free Society. (1945). *General Education in a Free Society: Report of the Harvard Committee*. Cambridge, MA: Harvard University Press. Available: https://isites.harvard.edu/fs/docs/icb.topic996234.files/general education032440mbp.pdf [July 2016].

Committee to Enquire into the Position of Natural Science in the Educational System of Great Britain. (1918). *Report of the Committee Appointed by the Prime Minister to Enquire into the Position of Natural Science in the Educational System of Great Britain*. London: Her Majesty's Stationary Office. Available: https://archive.org/details/reportofcommitte00greauoft [July 2016].

Conrad, C.C., and Hilchey, K.G. (2011). A review of citizen science and community-based environmental monitoring: Issues and opportunities. *Environmental Monitoring and Assessment, 176*(1-4), 273-291.

Cooper-Robbins, S.C., Bernard, D., McCaffery, K., Brotherton, J.M., and Skinner, S.R. (2010). "I just signed": Factors influencing decision-making for school-based HPV vaccination of adolescent girls. *Health Psychology, 29*(6), 618-625.

Corburn, J. (2007). Community knowledge in environmental health science: Co-producing policy expertise. *Environmental Science & Policy, 10*(2), pp. 150-161.

Cossons, N. (1993). Let us take science into our culture. *Interdisciplinary Science Reviews, 18*(4), 337-342.

Council of Canadian Academies. (2014). *Science Culture: Where Canada Stands: Expert Panel on the State of Canada's Science Culture*. Ottawa, Canada: Council of Canadian Academies.

Crowell, A., and Schunn, C. (2015). Unpacking the relationship between science education and applied scientific literacy. *Research in Science Education, 46*(1), 129-140.

Dainton, F.S. (1968). *The Dainton Report: An Inquiry into the Flow of Candidates into Science and Technology*. London: Her Majesty's Stationary Office.

Davis, I.C. (1935). The measurement of scientific attitudes. *Science Education, 19*(3), 117-122.

Davis, R.C. (1958). *The Public Impact of Science in the Mass Media: A Report on a Nationwide Survey for the National Association of Science Writers*. Ann Arbor, MI: Inter-university Consortium for Political and Social Research.

Davis, T.C., Long, S.W., Jackson, R.H., Mayeaux, E.J., George, R.B., Murphy, P.W., and Crouch, M.A. (1993). Rapid estimate of adult literacy in medicine: A shortened screening instrument. *Family Medicine, 25*(6), 391-395.

Davis, T.C., Wolf, M., Bass, P., Middlebrooks, M., Kennen, E., Baker, D., Bennett, C., Durazo-Arvizu, R., Bocchini, A., Savory, S., and Parker, R. (2006). Low literacy impairs comprehension of prescription drug warning labels. *Journal of General Internal Medicine, 21*(8), 847-851.

Dawson, E. (2014). "Not designed for us": How science museums and science centers socially exclude low-income, minority ethnic groups. *Science Education, 98*(6), 981-1008.

DeBoer, G.E. (2000). Scientific literacy: Another look at its historical and contemporary meanings and its relationship to science education reform. *Journal of Research in Science Teaching, 37*(6), 582-601.

Delgado, R., and Stefancic, J. (2012). *Critical Race Theory: An Introduction*. New York: New York University Press.

Delli Carpini, M., Cook, F., and Jacobs, L. (2004). Public deliberation, discursive participation, and citizen engagement: A review of the empirical literature. *Annual Review of Political Science, 7*, 315-344.

DeVellis, R.F. (2003). *Scale Development: Theory and Applications* (2nd ed.). Thousand Oaks, CA: SAGE.

DeWalt, D., and Hink, A. (2009). Health literacy and child health outcomes: A systematic review of the literature. *Pediatrics, 124*(3), S265-S274.

DeWalt, D., Berkman, N., Sheridan, S., Lohr, K., and Pignone, M. (2004). Literacy and health outcomes: A systematic review of the literature. *Journal of General Internal Medicine, 19*(12), 1228-1239.

DeWalt, D., Malone, R., Bryant, M., Kosnar, M., Corr, K., Rothman, R., Sueta, C., and Pignone, M. (2006). A heart failure self-management program for patients of all literacy levels: A randomized, controlled trial. *BMC Health Services Research, 6*, 30.

DeWalt, D., Boone, R., and Pignone, M. (2007). Literacy and its relationship with self-efficacy, trust, and participation in medical decision making. *American Journal of Health Behavior, 31*(Suppl. 1), S27-S35.

Dewey, J. (1927). *The Public and Its Problems: An Essay in Political Inquiry*. New York: Holt.

Dewey, J. (1934). The supreme intellectual obligation. *Science Education, 18*(1), 1-4.

Doble, J. (1995). Public opinion about issues characterized by technological complexity and scientific uncertainty. *Public Understanding of Science, 4*(2), 95-118.

Drew, J.A. (2005). Use of traditional ecological knowledge in marine conservation. *Conservation Biology, 19*(4), 1286-1293.

Druckman, J., and Bolsen, T. (2011). Framing, motivated reasoning, and opinions about emergent technologies. *Journal of Communication, 61*(4), 659-688.

Drummond, C., and Fischhoff, B. (2015). Development and validation of the scientific reasoning scale. *Journal of Behavioral Decision Making*, e1-e13.

Dudo, A., Brossard, D., Shanahan, J., Scheufele, D., Morgan, M., and Signorielli, N. (2010). Science on television in the 21st century: Recent trends in portrayals and their contributions to public attitudes toward science. *Communication Research, 38*(6), 754-777.

Dunn, A.G., Leask, J., Zhou, X., Mandl, K.D., and Coiera, E. (2015). Associations between exposure to and expression of negative opinions about human papillomavirus vaccines on social media: An observational study. *Journal of Medical Internet Research, 17*(6), e144.

Durant, J.R., Evans, G.A., and Thomas, G.P. (1989). The public understanding of science. *Nature, 340*(6228), 11-14.

Durant, J., Bauer, M., Midden, C., Gaskell, G., and Liakopoulos, M. (2000). Two cultures of public understanding of science. In M. Dierkes and C.V. Grote (Eds.), *Between Understanding and Trust: The Public, Science, and Technology* (pp. 131-156). Amsterdam: Harwood Academic.

Einsiedel, E., and Eastlick, D. (2000). Consensus conferences as deliberative democracy: A communications perspective. *Science Communication, 21*(4), 323-343.

Ellis, J., Mullan, J., Worsley, A., and Pai, N. (2012). The role of health literacy and social networks in arthritis patients' health information-seeking behavior: A qualitative study. *International Journal of Family Medicine, 1*(Art. 397039), 1-6.

Epstein, S. (1995). The construction of lay expertise: AIDS activism and the forging of credibility in the reform of clinical trials. *Science, Technology & Human Values, 20*(4), 408-437.

Epstein, S. (1996). *Impure Science: AIDS, Activism, and the Politics of Knowledge*. Berkeley: University of California Press.

Epstein, S. (2008). Patient groups and health movements. In E.J. Hackett, O. Amsterdamska, M. Lynch, and J. Wajcman (Eds.), *The Handbook of Science and Technology Studies* (pp. 499-539). Cambridge, MA: MIT Press.

European Commission. (1995). *White Paper on Education and Training*. Brussels, Belgium: European Commission.

European Commission. (1997). *European Opinions on Modern Biotechnology: Eurobarometer 46.1*. Brussels, Belgium: European Commission. Available: http://ec.europa.eu/public_opinion/archives/ebs/ebs_108_en.pdf [July 2016].

European Commission. (2004). *Europe Needs More Scientists: Report by the High-Level Group on Increasing Human Resources for Science and Technology*. Brussels, Belgium: European Commission.

European Commission. (2013). *Special Eurobarometer 401: Responsible Research and Innovation (RRI), Science and Technology*. Brussels, Belgium: European Commission. Available: http://ec.europa.eu/public_opinion/archives/ebs/ebs_401_en.pdf [June 2016].

Evans, C., Abrams, E., Reltsrna, R., Roux, K., Salmonsen, L., and Marra, P.P. (2005). The neighborhood nest-watch program: Participant outcomes of a citizen-science ecological research project. *Conservation Biology, 19*(3), 589-594.

Evans, G., and Durant, J. (1995). The relationship between knowledge and attitudes in the public understanding of science in Britain. *Public Understanding of Science, 4*(1), 57-74.

Feinstein, N.W. (2011). Salvaging science literacy. *Science Education, 95*(1), 168-185.

Feinstein, N.W., Allen, S., and Jenkins, E. (2013). Outside the pipeline: Reimagining science education for nonscientists. *Science, 340*(6130), 314-317.

Feldman, L., and Hart, P. (2016). Using policitcal efficacy messages to increase climate activism: The mediating role of emotions. *Science Communication, 38*(1), 99-127.

Fessenden-Raden, J., Fitchen, J., and Heath, J. (1987). Providing risk information in communities: Factors influencing what is heard and accepted. *Science, Technology & Human Values, 12*(3/4), 94-101.

Finger, M. (1994). From knowledge to action? Exploring the relationships between environmental experiences, learning, and behavior. *Journal of Social Issues, 50*(3), 141-160.

Fiske, S., and Taylor, S. (1991). *Social Cognition* (2nd ed.). New York: McGraw-Hill.

Fourez, G. (1997). Scientific and technological literacy as social practice. *Social Studies of Science, 27*(6), 903-936.

Frank, M. (1989). Project 2061: Science for all Americans. *The Physiologist, 32*(5), 245-248.

Frickel, S., Gibbon, S., Howard, J., Kempner, J., Ottinger, G., and Hess, D. (2010). Undone science: Charting social movement and civil society challenges to research agenda setting. *Science, Technology, & Human Values, 35*(4), 444-473.

Frickel, S., Torcasso, R., and Anderson, A. (2015). The organization of expert activism: Shadow mobilization in two social movements. *Mobilization: An International Quarterly, 20*(3), 305-323.

Fung, T., Choi, D., Scheufele, D., and Shaw, B. (2014). Public opinion about biofuels: The interplay between party identification and risk/benefit perception. *Energy Policy, 73*, 344-355.

Funk, C., and Rainie, L. (2015). *Public and Scientists' Views on Science and Society*. Washington, DC: Pew Research Center. Available: http://www.pewinternet.org/2015/01/29/public-and-scientists-views-on-science-and-society [July 2016].

Funtowicz, S.O., and Ravetz, J.R. (1995). *Science for the Post Normal Age*. Dordrecht, Netherlands: Springer.

Gabel, L.L. (1976). *The Development of a Model to Determine Perceptions of Scientific Literacy*. Ph.D. Dissertation. Columbus: Ohio State University.

Gaskell, G., Allum, N.C., and Stares, S.R. (2003). *Europeans and Biotechnology in 2002*. Brussels, Belgium: Europa. Available: http://ec.europa.eu/public_opinion/archives/ebs/ebs_177_en.pdf [July 2016].

Gaskell, G., Allum, N.C., Wagner, W., Kronberger, N., Torgersen, H., Hampel, J., and Bardes, J. (2004). GM foods and the misperception of risk perception. *Risk Analysis, 24*(1), 185-194.

Gastil, J., and Dillard, J. (1999). Increasing political sophistication through public deliberation. *Political Communication, 16*(1), 3-23.

Gauchat, G. (2012). Politicization of science in the public sphere: A study of public trust in the United States, 1974 to 2010. *American Sociological Review, 77*(2), 167-187.

Gauchat, G. (2015). The political context of science in the United States: Public acceptance of evidence-based policy and science funding. *Social Forces, 94*(2), 723-746.

Ginsburg, L., Manly, M., and Schmitt, M.J. (2006). *The Components of Numeracy, Occasional Paper*. Boston, MA: National Center for the Study of Adult Learning and Literacy.

Glanz, K., and Rimer B. (1995). *Theory at a Glance: A Guide to Health Promotion Practice*. Bethesda, MD: National Cancer Institute.

Goidel, K., and Nisbet, M. (2006). Exploring the roots of public participation in the controversy over embryonic stem cell research and cloning. *Political Behavior, 28*(2), 175-192.

Golbeck, A.L., Ahlers-Schmidt, C., Paschal, A., and Dismuke, S.E. (2005). A definition and operational framework for health numeracy. *American Journal of Preventive Medicine, 29*(4), 375-376.

Goldacre, B. (2006). Bad science: How to put your relationship to the universe into a seasonal perspective. *The Guardian,* December 23. Available: https://www.theguardian.com/science/2006/dec/23/badscience.christmas2006 [July 2016].

Goldacre, B. (2008). *Bad Science*. London: Harper Collins.

Goldman, S., and Snow, C.E. (2015). Adolescent literacy: Development and instruction. In A. Pollatsek and R. Treiman (Eds.), *Handbook on Reading* (pp. 463-478). New York: Oxford University Press.

Goodman, M., Finnegan, R., Mohadjer, L., Krenzke, T., and Hogan, J. (2013). *Literacy, Numeracy, and Problem Solving in Technology-rich Environments among U.S. Adults: Results from the Program for the International Assessment of Adult Competencies 2012: First Look.* Washington, DC: National Center for Education Statistics, U.S. Department of Education.

Gore, P., Madhavan, S., Curry, D., McClung, G., Castigilia, M., Rosenbluth, S., and Smego, R. (1999). Predictors of childhood immunization completion in a rural population. *Social Science & Medicine, 48*(8), 1011-1027.

Greenberg, M., and Wartenberg, D. (1991). Communicating to an alarmed community about cancer clusters: A fifty state study. *Journal of Community Health, 16*(2), 71-82.

Gunther, A., and Liebhart, J. (2006). Broad reach or biased source? Decomposing the hostile media effect. *Journal of Communication, 56*(3), 449-466.

Guy, S., Kashima, Y., Walker, I., and O'Neill, S. (2014). Investigating the effects of knowledge and ideology on climate change beliefs. *European Journal of Social Psychology, 44*(5), 421-429.

Hanushek, E., and Woessman, L. (2016). Knowledge capital, growth, and the East Asian miracle. *Science, 351*(6271), 344-345.

Hargreaves, I., and Ferguson, G. (2000) *Who's Misunderstanding Whom? Science, Society, and the Media.* Swindon, UK: Economic and Social Research Council.

Harrison, J.L. (2011). Parsing "participation" in action research: Navigating the challenges of lay involvement in technically complex participatory science projects. *Society and Natural Resources, 24*(7), 702-716.

Hart, P.S., and Nisbet, E.C. (2012). Boomerang effects in science communication: How motivated reasoning and identity cues amplify opinion polarization about climate mitigation policies. *Communication Research, 39*, 701-723.

Hart, P.S., Nisbet, E.C., and Myers, T.A. (2015). Public attention to science and political news and support for climate change mitigation. *Nature Climate Change, 5*, 541-545.

Haun, J.N., Valerio, M.A., McCormack, L.A., Sørensen, K., and Paasche-Orlow, M.K. (2014). Health literacy measurement: An inventory and descriptive summary of 51 instruments. *Journal of Health Communication, 19*(Suppl. 2), 302-333.

Hawthorne, G. (1997). Preteenage drug use in Australia: The key predictors and school-based drug education. *Journal of Adolescent Health, 20*(5), 384-395.

Henderson, B., Osborne, J., MacPherson, A., and Wild, A. (2015). Beyond construction: Five arguments for the role of critique in teaching science. *International Journal of Science Education, 37*(10), 1668-1697.

Hess, D., Breyman, S., Campbell, N., and Martin, B. (2008). Science, technology, and social movements. In E.J. Hackett, O. Amsterdamska, M. Lynch, and J. Wajcman (Eds.), *The Handbook of Science and Technology Studies* (pp. 473-498). Cambridge, MA: MIT Press.

Hines, J., Hungerford, H., and Tomera, A. (1986). Analysis and synthesis of research on responsible environmental behavior: A meta-analysis. *Journal of Environmental Education, 18*(2), 1-8.

Hirsch, E.D. (1987). *Cultural Literacy: What Every American Needs to Know.* Boston, MA: Houghton Mifflin.

Hirst, P.H. (1965). Liberal education and the nature of knowledge. *Philosophical Analysis and Education, 2*, 113-140.

Hirst, P.H., and Peters, R.S. (1970). *The Logic of Education.* London: Routledge.

Ho, S.S., Brossard, D., and Scheufele, D.A. (2008). Effects of value predispositions, mass media use, and knowledge on public attitudes toward embryonic stem cell research. *International Journal of Public Opinion Research, 20*(2), 171-192.

Ho, S.S., Scheufele, D.A., and Corley, E.A. (2010). Making sense of policy choices: Understanding the roles of value predispositions, mass media, and cognitive processing in public attitudes toward nanotechnology. *Journal of Nanoparticle Research, 12*(8), 2703-2715.

Ho, S.S., Scheufele, D.A., and Corley, E.A. (2011). Factors influencing public risk-benefit considerations of nanotechnology: Assessing the effects of mass media, interpersonal communication, and elaborative processing. *Public Understanding of Science, 22*(5), 606-623.

Hsu, S.J. (2004). The effects of an environmental education program on responsible environmental behavior and associated environmental literacy variables in Taiwanese college students. *The Journal of Environmental Education, 35*(2), 37-48.

Huntington, H. (2000). Using traditional ecological knowledge in science: Methods and applications. *Ecological Applications*, *10*(5), 1270-1274.

Hurd, P. (1958). Science literacy: Its meaning for American schools. *Educational Leadership*, *16*, 13-16.

Institute of Medicine. (2004). *Health Literacy: A Prescription to End Confusion*. L. Nielsen-Bohlman, A.M. Panzer, and D.A. Kindig (Eds.). Committee on Health Literacy, Board on Neuroscience and Behavioral Health. Washington, DC: The National Academies Press.

Institute of Medicine. (2011). *Innovations in Health Literacy Research: Workshop Summary*. C. Vancheri (Rapporteur). Roundtable on Health Literacy, Board on Population Health and Public Health Practice. Washington, DC: The National Academies Press.

Institute of Medicine. (2012). *How Can Health Care Organizations Become More Health Literate? Workshop Summary*. Washington, DC: The National Academies Press. doi:10.17226/13402.

Irwin, A., and Wynne, B. (Eds.). (1996). *Misunderstanding Science?: The Public Reconstruction of Science and Technology*. Cambridge, UK: Cambridge University Press.

Israel, B.A., Parker, E.A., Rowe, Z., Salvatore, A., Minkler, M., López, J., and Potito, P.A. (2005). Community-based participatory research: Lessons learned from the Centers for Children's Environmental Health and Disease Prevention Research. *Environmental Health Perspectives*, *113*(10), 1463-1471.

Jasanoff, S. (2003). Technologies of humility: Citizen participation in governing science. *Minerva*, *41*(3), 223-244.

Jasanoff, S. (2009). *Science at the Bar: Law, Science, and Technology in America*. Cambridge, MA: Harvard University Press.

Jastak, S., and Wilkinson, G. (1984). *Wide Range Achievement Test-Revised (WRAT-R)*. San Antonio, TX: The Psychological Corporation.

Jayanti, R., and Burns, A. (1998). The antecedents of preventive health care behavior: An empirical study. *Journal of the Academy of Marketing Science*, *26*(1), 6-15.

Kahan, D.M. (2015). Climate-science communication and the measurement problem. *Political Psychology*, *36*, 1-43.

Kahan, D.M. (in press). "Ordinary science intelligence": A science-comprehension measure for study of risk and science communication, with notes on evolution and climate change. *Journal of Risk Research*.

Kahan, D.M., Braman, D., Slovic, P., Gastil, J., and Cohen, G. (2009). Cultural cognition of the risks and benefits of nanotechnology. *Nature Nanotechnology*, *4*(2), 87-90.

Kahan, D.M., Jenkins-Smith, H., and Braman, D. (2011). Cultural cognition of scientific consensus. *Journal of Risk Research*, *14*, 147-174.

Kahan, D.M., Peters, E., Wittlin, M., Slovic, P., Larrimore Ouellette, L., Braman, D., and Mandel, G. (2012). The polarizing impact of science literacy and numeracy on perceived climate change risks. *Nature Climate Change*, *2*(10), 732-735.

Kaplan, L. (2000). Public participation in nuclear facility decisions: Lessons from Hanford. In D.L. Kleinman (Ed.), *Science, Technology and Democracy* (pp. 67-83). Albany: State University of New York Press.

Kata, A (2010). A postmodern Pandora's box: Anti-vaccination misinformation on the Internet. *Vaccine*, 28, 1709-1716.

Keller, D., Wright, J., and Pace, H. (2008). Impact of health literacy on health outcomes in ambulatory care patients: A systemic review. *The Annals of Pharmacotherapy*, *42*(9), 1272-1281.

Kempton, W., Boster, J., and Hartley, J. (1995). *Environmental Values in American Culture*. Cambridge, MA: MIT Press.

Kickbusch, I.S. (2001). Health literacy: Addressing the health and education divide. *Health Promotion International*, *16*(3), 289-297.

Kinchy, A., Jalbert, K., and Lyons, J. (2014). What is volunteer water monitoring good for? Fracking and the plural logics of participatory science. *Political Power and Social Theory*, *27*(2), 259-289.

Kirsch, I.S. (2001). *The International Adult Literacy Survey (IALS): Understanding What Was Measured*. Princeton, NJ: Educational Testing Service. Available: https://www.ets.org/Media/Research/pdf/RR-01-25-Kirsch.pdf [July 2016].

Kirsch, I.S., Jungeblut, A., Jenkins, L., and Kolstad, A. (1993). *Adult Literacy in America: A First Look at the Findings of the National Adult Literacy Survey*. Washington, DC: National Center for Education Statistics, U.S. Department of Education.

Kitcher, P. (2010). The climate change debates. *Science, 328*(5983), 1230-1234.

Kleinman, D.L. (2000). Democratizations of science and technology. In D.L. Kleinman (Ed.) *Science, Technology, and Democracy* (pp. 139-166). Albany, NY: SUNY Press.

Kleinman, D.L., Kinchy, A.J., and Handelsman, J. (Eds.). (2005). *Controversies in Science and Technology: From Maize to Menopause* (vol. 1). Madison: University of Wisconsin Press.

Klopfer, L.E. (1969). *Case Histories and Science Education*. San Francisco, CA: Wadsworth.

Knight, A.J. (2009). Perceptions, knowledge and ethical concerns with GM foods and the GM process. *Public Understanding of Science*, *18*(2), 177-188.

Koeppen, K., Hartig, J., Klieme, E., and Leutner, D. (2008). Current issues in competence modeling and assessment. *Journal of Psychology*, *216*(2), 61-73.

Koh, H.K., and Rudd, R.E. (2015). The arc of health literacy. *Journal of the American Medical Association*, *314*(12), 1225-1226.

Kollmuss, A., and Agyeman, J. (2002). Mind the gap: Why do people act environmentally and what are the barriers to pro-environmental behavior? *Environmental Education Research*, *8*(3), 239-260.

Korean Foundation for the Advancement of Science and Creativity (2013). *Survey of Public Attitudes towards and Understanding of Science and Technology 2012*. Seoul, Korea: KOFAC.

Kronberger, N., Holtz, P., and Wagner, W. (2012). Consequences of media information uptake and deliberation: Focus groups' symbolic coping with synthetic biology. *Public Understanding of Science*, *21*(2), 174-187.

Ladson-Billings, G. (2006). From the achievement gap to the education debt: Understanding achievement in US schools. *Educational Researcher*, *35*(7), 3-12.

Ladwig, P., Dalrymple, K.E., Brossard, D., Scheufele, D.A., and Corley, E.A. (2012). Perceived familiarity or factual knowledge? Comparing operationalizations of scientific understanding. *Science and Public Policy*, *39*(6), 761-774.

Lakshminarayanan, S. (2007). Using citizens to do science versus citizens as scientists. *Ecology and Society*, *12*(2), 2.

Laugksch, R.C. (2000). Scientific literacy: A conceptual overview. *Science Education*, *84*(1), 71-94.

Laugksch, R.C., and Spargo, P.E. (1996). Construction of a paper-and-pencil test of basic scientific literacy based on selected literacy goals recommended by the American Association for the Advancement of Science. *Public Understanding of Science*, *5*(4), 331-359.

Lave, J., and Wenger, E. (1991). *Situated Learning: Legitimate Peripheral Participation*. New York: Cambridge University Press.

Layton, D., Jenkins, E., Macgill, S., and Davey, A. (1993). *Inarticulate Science? Perspectives on the Public Understanding of Science and Some Implications for Science Education*. Nafferton, UK: Studies in Education.

Layton, D., Jenkins, E., and Donnelly, J. (1994). Scientific and technological literacy: Meanings and rationales. In *Innovations in Science and Technology Education* (vol. VI). Paris, France: UNESCO.

Lederman, N.G. (2007). Nature of science: Past, present, and future. In S.K. Abell and N.G. Lederman (Eds.). *Handbook of Research on Science Education* (pp. 831-879). Mahwah, NJ: Lawrence Erlbaum Associates.

Lee, C.J., Scheufele, D.A., and Lewenstein, B.V. (2005). Public attitudes toward emerging technologies: Examining the interactive effects of cognitions and effect on public attitudes toward nanotechnology. *Science Communication*, *27*(2), 240-267.

Lee, S., and Roth, W.M. (2003). Science and the "good citizen": Community-based scientific literacy. *Science, Technology, and Human Values*, *28*(3), 403-424.

Lehrer, J. (2010). The truth wears off: Is there something wrong with the scientific method? *The New Yorker*, Dec. 13, 52-57.

Liang, X., Ho, S.S., Brossard, D., Scheufele, D.A., Xenos, M.A., Hao, X., and He, X. (2013). Value predispositions as perceptual filters: Comparing public attitudes toward nanotechnology in the United States and Singapore. *Public Understanding of Science*, *24*, 582-600.

Liang, X., Ho, S.S., Brossard, D., Xenos, M.A., Scheufele, D.A., Anderson, A.A., Hao, X., He, X. (2015). Value predispositions as perceptual filters: Comparing of public attitudes toward nanotechnology in the United States and Singapore. *Public Understanding of Science, 24*(5), 582-600.

Liu, X., Tang, S., and Bauer, M.W. (2012). Comparing the public understanding of science across China and Europe. In M.W. Bauer, R. Shukla, and N. Allum (Eds.), *The Culture of Science: How the Public Relates to Science Across the Globe*. New York: Routledge.

Lord Sainsbury of Turville. (2007). *The Race to the Top: A Review of Government's Science and Innovation Policies*. London: Her Majesty's Stationary Office. Available: http://www.rsc.org/images/sainsbury_review051007_tcm18-103118.pdf [July 2016].

Losh, S.C. (2010). Stereotypes about scientists over time among U.S. adults: 1983 and 2001. *Public Understanding of Science*, *19*(3), 372-382.

Lowell, B.L., and Salzman, H. (2007). *Into the Eye of the Storm: Assessing the Evidence on Science and Engineering Education, Quality, and Workforce Demand*. Washington, DC: Urban Institute.

Machtinger, E., Wang, F., Chen, L., Rodriguez, M., Wu, S., and Schillinger, D. (2007). A visual medication schedule to improve anticoagulation control: A randomized, controlled trial. *Joint Commission Journal on Quality and Patient Safety*, *33*(10), 625-635.

Margolis, J., Estrella, R., Goode, J., Holme, J.J., and Nao, K. (2010). *Stuck in the Shallow End: Education, Race, and Computing*. Cambridge, MA: MIT Press.

Martin, M.O., Mullis, V.S., Gonzalez, E.J., Smith, T., Chrostowski, S., Garden, R., and O'Connor, K. (2000). *TIMSS 1999: International Science Report: Findings from IEA's Repeat of the Third International Mathematics and Science Study at Eighth Grade*. Boston, MA: International Association for the Evaluation of Educational Achievement.

Martin, M.O., Mullis, I., Gonalez, E.J., and Chrostowski, S. (2004). *TIMSS 2003: International Mathematics Report: Findings from IEA's Trends in International Mathematics and Science Study at the Fourth and Eighth Grades*. Boston, MA: International Association for the Evaluation of Educational Achievement.

Martin, M.O., Mullis, I., and Foy, P. (2008). *TIMSS 2007: International Science Report: Findings from IEA's Trends in International Mathematics and Science Study at the Fourth and Eighth Grades*. Boston, MA: International Association for the Evaluation of Educational Achievement.

Martin, M.O., Mullis, I., Foy, P., and Stanco, G. (2012). *TIMSS 2011 International Results in Science*. Boston, MA: International Association for the Evaluation of Educational Achievement.

McComas, K., Besley, J.C., and Steinhardt, J. (2014). Factors influencing U.S. consumer support for genetic modification to prevent crop disease. *Appetite*, *78*(1), 8-14.

McCurdy, R. (1958). Toward a population literate in science. *The Science Teacher*, *25*, 366-368.

Mercier, H., and Sperber, D. (2011). Why do humans reason? Arguments for an argumentative theory. *Behavioral and Brain Sciences*, *34*(2), 57-74.

Merton, R.K. (1968). The Matthew effect in science. *Science*, *159*(3810), 56-63.

Midden, C., Boy, D., Einsiedel, E., Fjoestad, B., Liakopoulos, M., Miller, J.D., and Wagner, W. (2002). The structure of public perceptions. In M.W. Bauer and G. Gaskell (Eds.), *Biotechnology: The Making of a Great Controversy* (pp. 203-223). Cambridge, MA: Cambridge University Press.

Millar, R., and Osborne, J.F. (Eds.). (1998). *Beyond 2000: Science Education for the Future*. London: King's College London.

Miller, J.D. (1983). Scientific literacy: A conceptual and empirical review. *Daedalus, 112*(2), 29-48.

Miller, J.D. (1998). The measurement of scientific literacy. *Public Understanding of Science, 7*(1), 203-223.

Miller, J.D. (2004). Public understanding of and attitudes toward scientific research: What we know and what we need to know. *Public Understanding of Science, 13*(3), 273-294.

Miller, J.D. (2010). Adult science learning in the Internet era. *Curator, 53*(2), 191-208.

Minkler, M., Vásquez, V.B., Tajik, M., and Petersen, D. (2008). Promoting environmental justice through community-based participatory research: The role of community and partnership capacity. *Health Education & Behavior, 35*(1), 119-137.

Moore, K. (2006). Powered by the people: Scientific authority in participatory science. In S. Frickel and K. Moore (Eds.), *The New Political Sociology of Science: Institutions, Networks, and Power* (pp. 299-323). Madison: University of Wisconsin Press.

Morris, N., MacLean, C., and Littenberg, B. (2006). Literacy and health outcomes: A cross-sectional study in 1002 adults with diabetes. *BMC Family Practice, 7*, 49.

Muñoz, A., Moreno, C., and Luján, J.L. (2012). Who is willing to pay for science? On the relationship between public perception of science and the attitude to public funding of science. *Public Understanding of Science, 21*(2), 242-253.

National Academy of Sciences, National Academy of Engineering, and Institute of Medicine. (2007). *Rising Above the Gathering Storm: Energizing and Employing America for a Brighter Economic Future*. Committee on Prospering in the Global Economy of the 21st Century: An Agenda for American Science and Technology; Committee on Science, Engineering, and Public Policy. Washington, DC: The National Academies Press.

National Academy of Sciences, National Academy of Engineering, and Institute of Medicine. (2010). *Rising Above the Gathering Storm Revisited: Rapidly Approaching Category 5*. Washington, DC: The National Academies Press.

National Assessment of Adult Literacy. (2003). *National Assessment of Adult Literacy (NAAL)*. Washington, DC: The National Center for Education Statistics. Available: https://nces.ed.gov/naal [July 2016].

National Center for Education Statistics. (2003). *Adult Literacy and Lifeskills Survey*. Washington, DC: National Center for Education Statistics. Available: https://nces.ed.gov/surveys/all [July 2016].

National Center for Education Statistics. (2012a). *Program for the International Assessment of Adult Competencies (PIAAC): Literacy Domain*. Washington, DC: National Center for Education Statistics. Available: https://nces.ed.gov/surveys/piaac/literacy.asp [July 2016].

National Center for Education Statistics. (2012b). *The Nation's Report Card: Science 2011*. Washington, DC: Institute of Education Sciences, U.S. Department of Education. Available: http://nces.ed.gov/nationsreportcard/pdf/main2011/2012465.pdf [June 2016].

National Institute of Science and Technology Policy. (2012). *Research on Changes in Public Awareness of Science and Technology: The Results of Interviews and Monthly Internet Surveys*. Tokyo, Japan: Ministry of Education.

National Research Council. (1996). *National Science Education Standards*. Washington, DC: National Academy Press.

National Research Council. (2009). *Strengthening Forensic Science in the United States: A Path Forward*. Committee on Identifying the Needs of the Forensic Sciences Community; Committee on Science, Technology, and Law; Policy and Global Affairs; Committee on Applied and Theoretical Statistics; Division on Engineering and Physical Sciences. Washington, DC: The National Academies Press.

National Research Council. (2012). *A Framework for K-12 Science Education: Practices, Crosscutting Concepts, and Core Ideas*. Committee on a Conceptual Framework for New K-12 Science Education Standards, Board on Science Education, Division of Behavioral and Social Sciences and Education. Washington, DC: The National Academies Press.

National Research Council. (2015). *Trust and Confidence at the Interfaces of the Life Sciences and Society: Does the Public Trust Science? A Workshop Summary*. Board on Life Sciences, Division on Earth and Life Studies; Board on Science Education, Division of Behavioral and Social Sciences and Education. Washington, DC: The National Academies Press.

National Science Board. (1989). Public literacy and attitudes toward science and technology. In *Science and Engineering Indicators, 1989* (Ch. 8) (pp. 161-190). Arlington, VA: National Science Foundation.

National Science Board. (1996). Science and technology: Public attitudes and public understanding. In *Science and Engineering Indicators, 1996* (Ch. 7, pp. 7.1-7.21). Arlington, VA: National Science Foundation.

National Science Board. (2008). *Science and Engineering Indicators 2008*. Arlington, VA: National Science Foundation.

National Science Board. (2012). Science and technology: Public attitudes and public understanding. In *Science and Engineering Indicators, 2012* (Ch. 7, pp. 7.1-7.51). Arlington, VA: National Science Foundation.

National Science Board. (2016). *Science and Engineering Indicators, 2016*. Arlington, VA: National Science Foundation.

Nisbet, M.C. (2005). The competition for worldviews: Values, information, and public support for stem cell research. *International Journal of Public Opinion Research, 17*(1), 90-112.

Nisbet, M.C., and Goidel, R. (2007). Understanding citizen perceptions of science controversy: Bridging the ethnographic-survey research divide. *Public Understanding of Science, 16*(4), 421-440.

Nisbet, M.C., Scheufele, D., Shanahan, J., Moy, P., Brossard, D., and Lewenstein, B. (2002). Knowledge, reservation, or promise? A media effects model for public perceptions of science and technology. *Communication Research, 29*(5), 584-608.

Norris, S.P. (1995). Learning to live with scientific expertise: Toward a theory of intellectual communalism for guiding science teaching. *Science Education, 79*(2), 201-217.

Norris, S.P. (1997). Intellectual independence for nonscientists and other content-transcendent goals of science education. *Science Education, 81*(2), 239-258.

Norris, S.P., and Phillips, L. (2003). How literacy in its fundamental sense is central to scientific literacy. *Science Education, 87*(2), 224-240.

Norris, S.P., Phillips, L., and Burns, D. (2014). Conceptions of scientific literacy: Identifying and evaluating their programmatic elements. In M. Matthews (Ed.), *International Handbook of Research in History, Philosophy and Science Teaching* (pp. 1317-1344). Dordrecht, Netherlands: Springer.

Nussbaum, M.C. (2000). *Women and Human Development: The Capabilities Approach*. Cambridge, UK: Press Syndicate of the University of Cambridge.

Nutbeam, D. (2008). The evolving concept of health literacy. *Social Science Medicine, 67*(12), 2072-2078.

Nyhan, B., Reifler, J., Richey, S., and Freed, G. (2014). Effective messages in vaccine promotion: A randomized trial. *Pediatrics, 133*(4), e835-e842.

O'Connor, R.E., Bord, R.J., and Fisher, A. (1999). Risk perceptions, general environmental beliefs, and willingness to address climate change. *Risk Analysis, 19*(3), 461-471.

OECD. (2006). *The PISA 2006 Assessment Framework for Science, Reading, and Mathematics.* Paris, France: OECD.

OECD. (2007). *PISA 2006: Science Competencies for Tomorrow's World* (vol. 1). Paris, France: OECD.

OECD. (2009). *The PISA 2009 Assessment Framework: Key Competencies in Reading, Mathematics, and Science.* Available: http://www.oecd.org/pisa/pisaproducts/44455820.pdf [July 2016].

OECD. (2012a). *Assessment and Analytical Framework.* Available: http://www.oecd.org/pisa/pisaproducts/PISA%202012%20framework%20e-book_final.pdf [July 2016].

OECD. (2012b). *The PISA 2012 Mathematics Framework.* Available: http://www.oecd.org/pisa/pisaproducts/46961598.pdf [July 2016].

OECD. (2013). *The PISA 2015 Draft Science Framework.* Available: http://www.oecd.org/pisa/pisaproducts/Draft%20PISA%202015%20Science%20Framework%20.pdf [July 2016].

OECD. (2014). *PISA 2012 Results: What Students Know and Can Do: Student Performance in Mathematics, Reading, and Science.* (Vol. I., Revised edition, February 2014). Paris, France: OECD.

OECD and UNESCO Institute for Statistics (2003). *Literacy Skills for the World of Tomorrow – Further results from PISA 2000.* Paris, France: OECD.

O'Rourke, D., and Macey, G.P. (2003). Community environmental policing: Assessing new strategies of public participation in environmental regulation. *Journal of Policy Analysis and Management, 22*(3), 383-414.

Osborn, C., Paasche-Orlow, M., Davis, T., and Wolf, M. (2007). Health literacy: An overlooked factor in understanding HIV health disparities. *American Journal of Preventive Medicine, 33*(5), 374-378.

Osborn, C., Paasche-Orlow, M., Cooper Bailey, S., and Wolf, M. (2011). The mechanisms linking health literacy to behavior and health status. *American Journal of Health Behavior, 35*(1), 118-128.

Osborne, J., and Dillon, J. (2008). *Science Education in Europe: Critical Reflections* (vol. 13). London: Nuffield Foundation.

Ottinger, G. (2010). Buckets of resistance: Standards and effectiveness of citizen science. *Science, Technology, and Human Values, 35*(2), 244-270.

Ottinger, G. (2013). *Refining Expertise: How Responsible Engineers Subvert Environmental Justice Challenges.* New York: New York University Press.

Ottinger, G. and Cohen, B.R. (2011). *Technoscience and Environmental Justice Expert Cultures in a Grassroots Movement.* Cambridge, MA: MIT Press.

Ownby, R.L., Acevedo, A., Waldrop-Valverde, D., Jacobs, R.J., and Caballero, J. (2014). Abilities, skills and knowledge in measures of health literacy. *Patient Education and Counseling, 95*(2), 211-217.

Paasche-Orlow, M., and Wolf, M. (2007). The causal pathways linking health literacy to health outcomes. *American Journal of Health Behavior, 31*(Suppl. 1), S19-S26.

Paasche-Orlow, M., Riekert, K., Bilderback, A., Chanmugam, A., Hill, P., Rand, C., Brancati, F., and Krishnan, J. (2005). Tailored education may reduce health disparities in asthma self-management. *American Journal of Respiratory and Critical Care Medicine, 172*(8), 980-986.

Paasche-Orlow, M., Wilson, E., and McCormack, L. (2010) The evolving field of health literacy research. *Journal of Health Communication, 15*(Suppl. 2), 5-8.

Pardo, R., and Calvo, F. (2002). Attitudes toward science among the European public: A methodological analysis. *Public Understanding of Science, 11*(2), 155-195.

Pardo, R., and Calvo, F. (2004). The cognitive dimension of public perceptions of science: Methodological issues. *Public Understanding of Science, 13*(3), 203-227.

Parker, R.M., Baker, D.W., Williams, M.V., and Nurss, J.R. (1995). The test of functional health literacy in adults: A new instrument for measuring patients' literacy skills. *Journal of General Internal Medicine, 10*(10), 537-541.

Pelikan, J., Röthlin, F., and Ganahl, K. (2012). *Comparative Report on Health Literacy in Eight EU Member States*. Available: http://www.maastrichtuniversity.nl/web/Institutes/FHML/CAPHRI/DepartmentsCAPHRI/InternationalHealth/ResearchINTHEALTH/Projects/HealthLiteracyHLSEU/MeasuringHealthLiteracyInEurope.htm [July 2016].

Pella, M.O., O'Hearn, G.T., and Gale, C.W. (1966). Referents to scientific literacy. *Journal of Research in Science Teaching, 4*(3), 199-208.

Perrin, A., and Duggan, M. (2015). *Americans' Internet Access: 2000-2015.* Washington, DC: Pew Research Center. Available: http://www.pewinternet.org/files/2015/06/2015-06-26_internet-usage-across-demographics-discover_FINAL.pdf [September 2016].

Peters, E. (2012). Beyond comprehension: The role of numeracy in judgments and decisions. *Current Directions in Psychological Science, 21*(1), 31-35.

Pew Research Center. (2015). *A Look at What the Public Knows and Does Not Know About Science*. Available: http://www.pewinternet.org/files/2015/09/2015-09-10_science-knowledge_FINAL.pdf [July 2016].

Phadke, R. (2005). People's science in action: The politics of protest and knowledge brokering in India. *Society and Natural Resources, 18*(4), 363-375.

Pleasant, A., Rudd, R., O'Leary, C., Paasche-Orlow, M., Allen, M., Alvarado-Little, W., Myers, L., and Parson, K. (2016). *Considerations for a New Definition of Health Literacy*. Discussion Paper. Washington, DC: National Academy of Medicine. Available: http://nam.edu/considerations-for-a-new-definition-of-health-literacy [July 2016].

Plutzer, E. (2013). The racial gap in confidence in science: Explanations and implications. *Bulletin of Science Technology and Society, 33*(5-6), 146-157.

Poland, G.A., and Jacobson, R.M. (2001). Understanding those who do not understand: A brief review of the anti-vaccine movement. *Vaccine*, 19, 2440-2445.

Popper, K. (1963). *Conjectures and Refutations: The Growth of Scientific Knowledge*. London: Routledge and Kegan Paul.

Powers, B.J., Olsen, M.K., Oddone, E.Z., Thorpe, C.T., and Bosworth, H.B. (2008). Literacy and blood pressure—do healthcare systems influence this relationship? A cross-sectional study. *BMC Health Services Research, 8,* 219-228.

Pretty, J. (2003). Social capital and the collective management of resources. *Science, 302*(5652), 1912-1914.

Priest, S.H. (2001). Misplaced faith: Communication variables as predictors of encouragement for biotechnology development. *Science Communication, 23*(2), 97-110.

Priest, S.H., Bonfadelli, H., and Rusanen, M. (2003). The "trust gap" hypothesis: Predicting support for biotechnology across national cultures as a function of trust in actors. *Risk Analysis, 23*(4), 751-766.

Prochaska, J.O., DiClemente, C.C., and Norcross, J.C. (1992). In search of how people change. Applications to addictive behaviors. *Journal of American Psychology, 47*(9), 1102-1114.

Provasnik, S., Kastberg, D., Ferraro, D., Lemanski, N., Roey, S., and Jenkins, F. (2012). *Highlights from TIMSS 2011: Mathematics and Science Achievement of U.S. Fourth- and Eighth-Grade Students in an International Context* (NCES 2013-009 Revised). Washington, DC: National Center for Education Statistics, Institute of Education Sciences, U.S. Department of Education.

Rabinovich, A., Morton, T., and Birney, M. (2012). Communicating climate science: The role of perceived communicator's motives. *Journal of Environmental Psychology, 32*(1), 11-18.

Retzbach, J., Otto, L., and Maier, M. (2015). Measuring the perceived uncertainty of scientific evidence and its relationship to engagement with science. *Public Understanding of Science* [Epub ahead of print].

Roberts, D.A. (2007). Scientific literacy/science literacy. In S. Abell and N.G. Lederman (Eds.), *Handbook of Research on Science Education* (pp. 729-780). Mahwah, NJ: Lawrence Erlbaum Associates.

Roberts, J., and Toffolon-Weiss, M. (2001). *Chronicles from the Environmental Justice Frontline*. New York: Cambridge University Press.

Robinson, L., Calmes, D., and Bazargan, M. (2008). The impact of literacy enhancement on asthma-related outcomes among underserved children. *Journal of the National Medical Association, 100*(8), 892-896.

Roduta Roberts, M., Reid, G., Schroeder, M., and Norris, S. (2013). Causal or spurious? The relationship of knowledge and attitudes to trust in science and technology. *Public Understanding of Science, 22*(5), 624-641.

Rogers, E.M. (1948). Science courses in general education. In E.J. McGrath (Ed.), *Science in General Education* (pp. 1-22). Dubuque, IA: W.C. Brown.

Ross, L., Frier, B., Kelnar, C., and Deary, I. (2001). Child and parental mental ability and glycemic control in children with Type 1 diabetes. *Diabetic Medicine, 18*(5), 364-369.

Ross, N., Medin, D., and Cox, D. (2007). Epistemological models and culture conflict: Menominee and Euro-American hunters in Wisconsin. *Ethos, 35*(4), 478-515.

Roth, W.M., and Calabrese Barton, A. (2004). *Rethinking Scientific Literacy.* New York: Routledge Falmer.

Roth, W.M., and Lee, S. (2002). Scientific literacy as collective praxis. *Public Understanding of Science, 11*(1), 33-56.

Roth, W.M., and Lee, S. (2004). Science education as/for participation in the community. *Science Education,* 24, 76-89.

Rothman, R., DeWalt, D., Malone, R., Bryant, B., Shintani, A., Crigler, B., Weinberger, M., and Pignone, M. (2004). Influence of patient literacy on the effectiveness of a primary care-based diabetes disease management program. *Journal of the American Medical Association, 292*(14), 1711-1716.

Rudd, R.E., Kirsch, I., and Yamamoto, K. (2004). *Literacy and Health in America.* Princeton, NJ: Educational Testing Service. Available: http://www.ets.org/Media/Research/pdf/PICHEATH.pdf [April 2016].

Rudd, R.E., McCray, A.T., and Nutbeam, D. (2012). Health literacy and definition of terms. In D. Begoray, D. Gillis, and G. Rowlands (Eds.), *Health Literacy in Context* (pp. 13-32). New York: Nova Science.

Rudolph, J.L. (2002). *Scientists in the Classroom: The Cold War Reconstruction of American Science Education* (1st Edition). New York: Palgave.

Rudolph, J.L. (2005). Epistemology for the masses: The origins of "the scientific method" in American schools. *History of Education Quarterly, 45*(3), 341-376.

Rudolph, J.L., and Horibe, S. (2015). What do we mean by science education for civic engagement? *Journal of Research in Science Teaching, 53*(6), 805-820.

Rutherford, F.J., and Ahlgren, A. (1989). *Science for All Americans: A Project 2061 Report.* Washington, DC: American Association for the Advancement of Science.

Ryder, J. (2001). Identifying science understanding for functional science literacy. *Studies in Science Education, 36*(1), 1-44.

Salzman, H. (2013). What shortages? The real evidence about the STEM workforce. *Issues in Science and Technology, 29*(4), 58-67.

Sanz-Menéndez, L., and Van Ryzin, G.G. (2013). Economic crisis and public attitudes toward science: A study of regional differences in Spain. *Public Understanding of Science,* 24(2), 167-182.

Sanz-Menéndez, L., Van Ryzin, G., and del Pino, E. (2014). Citizens' support for government spending on science and technology. *Science and Public Policy, 41*(5), 611-624.

Scheufele, D.A. (2006). *Messages and Heuristics: How Audiences Form Attitudes about Emerging Technologies.* London: The Wellcome Trust.

Scheufele, D.A. (2011). *Modern Citizenship or Policy Dead End? Evaluating the Need for Public Participation in Science Policy Making, and Why Public Meetings May Not Be the Answer.* Paper #R-34. Cambridge, MA: Harvard Kennedy School, Joan Shorenstein Center on Media, Politics, and Public Policy. Available: http://shorensteincenter.org/modern-citizenship-or-policy-dead-end-dietram-scheufele [July 2016].

Scheufele, D.A., and Lewenstein, B.V. (2005). The public and nanotechnology: How citizens make sense of emerging technologies. *Journal of Nanoparticle Research*, *7*(6), 659-667.

Scheufele, D.A., Corley, E.A., Shih, T.J., Dalrymple, K.E., and Ho, S.S. (2009). Religious beliefs and public attitudes toward nanotechnology in Europe and the United States. *Nature Nanotechnology*, *4*(2), 91-94.

Schillinger, D., Hammer, H., Wang, F., Palacios, J., McLean, I., Tang, A., Youmans, S., and Handley, M. (2008). Seeing in 3-D: Examining the reach of diabetes self-management support strategies in a public health care system. *Health Education Behavior*, *35*(5), 664-682.

Setälä, M., Grönlund, K., and Herne, K. (2010). Citizen deliberation on nuclear power: A comparison of two decision-making methods. *Political Studies*, *58*(4), 688-714.

Shamos, M.H. (1995). *The Myth of Scientific Literacy.* New Brunswick, NJ: Rutgers University Press.

Shen, B.S.P. (1975). Scientific literacy and the public understanding of science. In S.B. Day (Ed.), *Communication of Scientific Information* (pp. 44-52). Basel, Switzerland: Karger.

Shephard, K., Harraway, J., Lovelock, B., Skeaff, S., Slooten, L., Strack, M., Furnari, M., and Jowett, T. (2014). Is the environmental literacy of university students measurable? *Environmental Education Research*, *20*(4), 476-495.

Sheridan S., Halpern, D., Viera, A., Berkman, N., Donahue, K., and Crotty, K. (2011). Interventions for individuals with low health literacy: A systematic review. *Journal of Health Communication*, *16*(Suppl. 3), 30-54.

Shirk, J.L., Ballard, H.L., Wilderman, C.C., Phillips, T., Wiggins, A., Jordan, R., McCallie, E., Minarchek, M., Lewenstein, B.V., Krasny, M.E., and Bonney, R. (2012). Public participation in scientific research: A framework for intentional design. *Ecology and Society*, *17*(2), 29-48.

Siegel, H. (1988). *Educating Reason: Rationality, Critical Thinking and Education.* London: Routledge.

Sjøberg, L. (2002). Attitudes toward technology and risk: Going beyond what is immediately given. *Policy Sciences, 35*(4), 379-400.

Snow, C.E. (2002). *Reading for Understanding: Toward an R&D Program in Reading Comprehension.* Santa Monica, CA: RAND.

Snow, C.E. (2016). *Literacy.* Paper presented to Committee on Scientific Literacy and Public Perception of Science, March 7, National Academies of Sciences, Engineering, and Medicine, Washington, DC.

Sørensen, K, Van den Broucke, S., Fullam, J., Doyle, G., Pelikan, J., Slonska, Z., and Brand, H. (2012). Health literacy and public health: A systematic review and integration of definitions and models. *BMC Public Health*, *12*(80), 1-13.

Sørensen, K., Pelikan, J.M., Röthlin, F., Ganahl, K., Slonska, Z., Doyle, G., Fullam, J., Kondilis, B., Agrafiotis, D., Uiters, E., Falcon, M., Mensing, M., Tchamov, K., van den Broucke, S., and Brand, H. (2015). Health literacy in Europe: Comparative results of the European Health Literacy Survey (HLS-EU). *European Journal of Public Health* 25(6), 1053-1058.

Southwell, B.G. (2013). *Social Networks and Popular Understanding of Science and Health: Sharing Disparities.* Baltimore, MD: Johns Hopkins University Press.

Spencer, H. (1884). *What Knowledge Is of Most Worth.* New York: JB Alden.

Stanovich, K.E. (1986). Matthew effects in reading: Some consequences of individual differences in the acquisition of literacy. *Reading Research Quarterly*, *21*(4), 360-407.

Steen, L.A. (2001). *Mathematics and Democracy: The Case for Quantitative Literacy.* Princeton, NJ: National Council on Education and the Disciplines.

Sturgis, P., and Allum, N. (2004). Science in society: Re-evaluating the deficit model of public attitudes. *Public Understanding of Science*, *13*(1), 55-74.

Sturgis, P., and Allum, N.C. (2006). *A Literature Review of Research Conducted on Public Interest, Knowledge, and Attitudes to Biomedical Science.* Available: http://www.wellcome.ac.uk/stellent/groups/corporatesite/@msh_grants/documents/web_document/wtd038717.pdf [July 2016].

Sturgis, P., Brunton-Smith, I., and Fife-Schaw, C. (2010). Public attitudes to genomic science: An experiment in information provision. *Public Understanding of Science*, *19*(2), 166-180.

Sudore, R.L., and Schillinger, D. (2009). Interventions to improve care for patients with limited health literacy. *Journal of Clinical Outcomes Management*, *16*(1), 20-29.

Sugawara, Y., Narimatsu, H., Hozawa, A., Shao, L., Otani, K., and Fukao, A. (2012). Cancer patients on Twitter: A novel patient community on social media. *BMC Research Notes*, 5, 699. Available: http://bmcresnotes.biomedcentral.com/articles/10.1186/1756-0500-5-699 [September 2016].

Sum, A., Kirsch, I., and Taggart, R. (2002). *The Twin Challenges of Mediocrity and Inequality: Literacy in the U.S. from an International Perspective.* Princeton, NJ: Educational Testing Service.

The Royal Society. (2014). *Vision for Science and Mathematics Education.* London: The Royal Society.

Trochim, W. (2000). *The Research Methods Knowledge Base* (2nd ed.). Cincinnati, OH: Atomic Dog.

Turnbull, D. (1997). Reframing science and other local knowledge traditions. *Futures, 29*(6), 551-562.

Turnbull, D. (2003). *Masons, Tricksters, and Cartographers: Comparative Studies in the Sociology of Scientific and Indigenous Knowledge.* New York: Taylor & Francis.

Tytler, R., Duggan, S., and Gott, R. (2001). Public participation in an environmental dispute: Implications for science education. *Public Understanding of Science*, *10*(4), 343-364.

U.S. Department of Education. (2014a). *Civil Rights Data Collection Data Snapshot: College and Career Readiness* (Issue Brief Number 3). Washington, DC: Office for Civil Rights, U.S. Department of Education.

U.S. Department of Education. (2014b). *Expansive Survey of America's Public Schools Reveals Troubling Racial Disparities.* Available: http://www.ed.gov/news/press-releases/expansive-survey-americas-public-schools-reveals-troubling-racial-disparities [July 2016].

U.S. Department of Education, National Center for Education Statistics. (2004). *The Condition of Education 2004* (NCES 2004–077). Washington, DC: U.S. Government Printing Office.

Varkey, A., Manwell, L., Williams, E., Ibrahim, S., Brown, R., Bobula, J., Horner-Ibler, B., von Wagner, C., Steptoe, A., Wolf, M., and Wardle, J. (2009). Health literacy and health actions: A review and a framework from health psychology. *Health Education & Behavior*, *36*(5), 860-877.

Viswanath, K., and Finnegan, J.R. (1996). The knowledge gap hypothesis: Twenty-five years later. *Annals of the International Communication Association*, *19*(1), 187-228.

von Wagner, C., Steptoe, A., Wolf, M.S., and Wardle, J. (2009). Health literacy and health actions: A review and a framework from health psychology. *Health Education and Behavior, 36*(5), 860-877.

Wallerstein, N., and Duran, B. (2006). Using community-based participatory research to address health disparities. *Health Promotion Practice, 7*(3), 312-323.

Warschauer, M. (2006). *Laptops and Literacy: Learning in the Wireless Classroom.* New York: Teachers College Press.

Washington, S.H., and Pellow, D. (2016). Water crisis in Flint, Michigan: Interview with David Pellow, Ph.D. *Environmental Justice*, *9*(2), 53-58.

Weiss, B.D., Mays, M.Z., Martz, W., Castro, K.M., DeWalt, D.A., Pignone, M.P., Mockbee, J., and Hale, F.A. (2005). Quick assessment of literacy in primary care: The Newest Vital Sign. *Annals of Family Medicine*, *3*(6), 514-522.

Weissman, J. (2013). The Ph.D. bust: America's awful market for young scientists—in 7 charts. *Atlantic Monthly*, February 20.

Wellcome Trust. (2013). *Wellcome Trust Monitor: Wave 2.* Available: http://www.wellcome.ac.uk/stellent/groups/corporatesite/@msh_grants/documents/web_document/wtp053113.pdf [May 2015].

Weller, J.A., Dieckmann, N.F., Tusler, M., Mertz, C.K., Burns, W.J., and Peters, E. (2013). Development and testing of an abbreviated numeracy scale: A Rasch analysis approach. *Journal of Behavioral Decision Making*, *26*(2), 198-212.

Wenger, E. (2000). Communities of practice and social learning systems. *Organization*, *7*(2), 225-246.

Windschitl, M., Thompson, J., and Braaten, M. (2008). Beyond the scientific method: Model-based inquiry as a new paradigm of preference for school science investigations. *Science Education*, *92*(5), 941-967.

World Health Organization. (1998). *Health Promotion Glossary*. www.who.int/healthpromotion/about/HPR%20Glossary%201998.pdf [July 2016].

World Health Organization. (2009). *7th Global Conference on Health Promotion, Track 2: Health Literacy and Health Behavior*. http://www.who.int/healthpromotion/conferences/7gchp/track2/en [July 2016].

World Values Survey. (2014). *WVS Wave 6 2010-2014 Official Aggregate v.20150418*. Available: http://www.worldvaluessurvey.org/WVSDocumentationWV6.jsp [June 2016].

Wynne, B. (1992). Misunderstood misunderstanding: Social identities and public uptake of science. *Public Understanding of Science*, *1*(3), 281-304.

Xie, Y., and Killewald, A.A. (2012). *Is American Science in Decline?* Cambridge, MA: Harvard University Press.

Yeo, S., Cacciatore, M., and Scheufele, D. (2015). News selectivity and beyond: Motivated reasoning in a changing media environment. In O. Jandura, C. Mothes, T. Petersen, and A. Schielicke (Eds.), *Publizistik und gesellschaftliche Verantwortung: Festschrift für Wolfgang Donsbach* (pp. 83-104). Berlin, Germany: Verlag Springer.

Yin, H., Dreyer, B., van Schaick, L., Foltin, G., Dinglas, C., and Mendelsohn, A. (2008). Randomized controlled trial of a pictogram-based intervention to reduce liquid medication dosing errors and improve adherence among caregivers of young children. *Archives of Pediatrics and Adolescent Medicine*, *162*(9), 814-822.

Zarcadoolas, C., Pleasant, A., and Greer, D.S. (2005). Understanding health literacy: An expanded model. *Health Promotion International*, *20*(2), 195-203.

Zhang, N., Terry, A., and McHorney, C. (2014). Impact of health literacy on medication adherence: A systematic review and meta-analysis. *Annals of Pharmacotherapy*, *48*(6), 741-751.

Zwickle, A., Koontz, T.M., Slagle, K.M., and Bruskotter, J.T. (2014). Assessing sustainability knowledge of a student population: Developing a tool to measure knowledge in the environmental, economic and social domains. *International Journal of Sustainability in Higher Education*, *15*(4), 375-389.

Appendix A

Key Definitions and Statements about Literacy, Numeracy, Science Literacy, Health Literacy, and Health Numeracy

The five tables in this appendix detail the formal definitions and other definitional statements about literacy, numeracy (quantitative literacy), science literacy, health literacy, and health numeracy. The following acronyms are used in the tables:

AAAS	American Association for the Advancement of Science
ALL	Adult Literacy and Lifeskills Survey
AMA	American Medical Association
HALS	Health Activities Literacy Scale
IALS	International Adult Literacy Survey
IOM	Institute of Medicine
NAAL	National Assessment of Adult Literacy
NALS	National Adult Literacy Survey
NCES	National Center for Education Statistics
NRC	National Research Council
NSB	National Science Board
OECD	Organisation for Economic Co-operation and Development
PIAAC	Program for the International Assessment of Adult Competencies
PISA	Programme for International Student Assessment
REALM	Rapid Estimate of Adult Literacy in Medicine
S-TOFHLA	Short Test of Functional Health Literacy in Adults
TOFHLA	Test of Functional Health Literacy in Adults
WHO	World Health Organization

TABLE A-1 Literacy

Definition	Source	Assessment/ Educational Standard
The active engagement of the reader in constructing meaning through the accurate and fluent processing of text.	Snow (2002)	
Using printed and written information to function in society, to achieve one's goals, and to develop one's knowledge and potential.	Kirsch et al. (1993), Kirsch (2001)	NALS 1992, IALS 2001
How adults use printed and written information to adequately function at home, in the workplace, and in the community. Measures three types of literacy—**prose, document, and quantitative.**	NCES (2003)	NAAL 2003
The knowledge and skills needed to understand and use information from text and other written formats.	NCES (2003)	ALL 2003-2008
Reading literacy is understanding, using, reflecting on and engaging with written texts, in order to achieve one's goals, to develop one's knowledge and potential, and to participate in society. There are three broad aspect categories: (1) access and retrieve, (2) integrate and interpret, (3) reflect and evaluate.	OECD (2009)	PISA 2009
Understanding, evaluating, using and engaging with written text to participate in the society, to achieve one's goals and to develop one's knowledge and potential.	NCES (2012a)	PIAAC 2012
Definitions of literacy emphasize the active engagement of the reader in constructing meaning through the accurate and fluent processing of text and note that success at reading comprehension depends on language skills and world knowledge as well as on control over decoding processes. Task or purpose for literacy use is at the center of any interpretation of reader skill, with an emphasis on the sociocultural context in which literacy skills are deployed and the role of that context in determining what constitutes adequate literacy levels.	Snow (2016)	

TABLE A-2 Numeracy (quantitative literacy)

Definition	Source	Assessment/ Educational Standard
The knowledge and skills required to apply arithmetic operations, either alone or sequentially, to numbers embedded in printed materials, such as balancing a check-book, figuring out a tip, completing an order form, or determining the amount of interest on a loan from an advertisement.	Kirsch et al. (1993), Kirsch (2001)	NALS 1992, IALS 2001
To be numerate means to be competent, confident, and comfortable with one's judgements on *whether* to use mathematics in a particular situation and if so, *what* mathematics to use, *how* to do it, *what* degree of accuracy is appropriate, and *what* the answer means in relation to the context.	Coben (2000, 2003)	
A more comprehensive portrait of quantitative literacy includes (1) confidence with mathematics; (2) cultural appreciation of mathematics; (3) interpreting data; (4) logical thinking; (5) using mathematics in making decisions in everyday life; (6) using mathematics in specific settings; (7) number sense; (8) practical skills in wide variety of common situations; (9) prerequisite knowledge (ability to use algebraic, geometric, and statistical tools); (10) symbol sense (being comfortable with algebraic and other mathematical symbols).	Steen (2001)	
Numeracy is defined as the knowledge and skills required to manage mathematical demands of diverse situations.	NCES (2003)	ALL 2003-2008
The components of numeracy are (1) context—the use and purpose for which an adult takes on a task with mathematics demands; (2) content—the mathematical knowledge that is necessary for the tasks confronted; and (3) cognitive and affective—the processes that enable an individual to solve problems and, thereby, link the content and the context.	Ginsburg et al. (2006)	

continued

TABLE A-2 Continued

Definition	Source	Assessment/ Educational Standard
Mathematical literacy is an individual's capacity to formulate, employ, and interpret mathematics in a variety of contexts. It includes reasoning mathematically and using mathematical concepts, procedures, facts, and tools to describe, explain, and predict phenomena. It assists individuals to recognize the role that mathematics plays in the world and to make the well-founded judgments and decisions needed by constructive, engaged and reflective citizens.	OECD (2012b)	PISA 2012
Numeracy is the ability to use, apply, interpret, and communicate mathematical information and ideas. It is an essential skill in an age when individuals encounter an increasing amount and wide range of quantitative and mathematical information in their daily lives. Numeracy is a skill parallel to reading literacy, and it is important to assess how these competencies interact, since they are distributed differently across subgroups of the population.	Goodman et al. (2013), NCES (2012a)	PIAAC 2013
Numeracy is defined as the ability to access, use, interpret, and communicate mathematical information and ideas, to engage in and manage mathematical demands of a range of situations in adult life.	NCES (2012a)	PIAAC 2012
Numeracy is defined as the ability to understand probabilistic and mathematical concepts.	Peters (2012)	

TABLE A-3 Science Literacy

Definition	Source	Assessment/ Educational Standard
Attempts to define human values, to understand the social, economic and political problems of our times, or to validate educational objectives without a consideration of modern science are unrealistic. More than a casual acquaintance with scientific forces and phenomena is essential for effective citizenship today. Further efforts are required to choose learning experiences that have a particular value for the development of an appreciation of science as an intellectual achievement, as a procedure for exploration and discovery, and which illustrate the spirit of scientific endeavor.	Hurd (1958)	
Scientific literacy involves (1) an understanding of the basic concepts in science, The scientifically literate individual presently is characterized as one with and understanding of (a) the basic concepts in science, (b) the nature of science, (c) the ethics that control the scientist in his work, (d) the interrelationships of science and society, (e) the interrelationships of science and the humanities, and (f) the differences between science and technology.	Pella et al. (1966)	
Distinguish three forms of science literacy: • practical (scientific and technical know-how that can be immediately put to use to help improve living standards) • civic ([allows citizen to] participate more fully in the democratic processes of an increasingly technological society) • cultural (motivated by a desire to know something about science as a major human achievement)	Shen (1975a)	
The science-literate person is one who: is aware that science, mathematics and technology are interdependent human enterprises with strengths and limitations; understands key concepts and principles of science; is familiar with the natural world and recognizes both its diversity and unity; and uses scientific knowledge and scientific ways of thinking for individual and social purposes.	Frank (1989)	AAAS *Science for All Americans* 1989

continued

TABLE A-3 Continued

Definition	Source	Assessment/ Educational Standard
Reasons to care about the public understanding of science: (1) science is arguably the greatest achievement of our culture, and people deserve to know about it; (2) science affects everyone's lives, and people need to know about it; (3) many public policy decisions involve science, and these can only be genuinely democratic if they arise out of informed public debate; and (4) science is publicly supported, and such support is (or at least ought to be) based on at least a minimal level of public knowledge.	Durant et al. (1989)	
Four scales measuring (1) scientific interest, (2) factual scientific knowledge, (3) general attitudes to science, and (4) support for European Commission funded science.	Bauer et al. (1994)	Eurobarometer 1989
An education in science should contain at least three components: (a) learning science (the facts, laws, and theories of science); (b) learning about science (the philosophical, historical, and sociological foundations of science); and (c) learning to live with science. Students should be taught how to use criteria for judging experts: the role and weight of consensus; the role and weight of prestige in the scientific community; the role and weight of publication and successful competition for research grants; and so on... students need practice in judging the credibility of scientific experts. This practice should be based on real-world problems that currently affect their lives.	Norris (1995)	

TABLE A-3 Continued

Definition	Source	Assessment/ Educational Standard
Scientific literacy is the knowledge and understanding of scientific concepts and processes required for personal decision making, participation in civic and cultural affairs, and economic productivity. Scientific literacy means that a person can ask, find, or determine answers to questions derived from curiosity about everyday experiences. It means that a person has the ability to describe, explain, and predict natural phenomena. Scientific literacy entails being able to read with understanding articles about science in the popular press and to engage in social conversation about the validity of the conclusions. Scientific literacy implies that a person can identify scientific issues underlying national and local decisions and express positions that are scientifically and technologically informed. A literate citizen should be able to evaluate the quality of scientific information on the basis of its source and the methods used to generate it. Scientific literacy also implies the capacity to pose and evaluate arguments based on evidence and to apply conclusions from such arguments appropriately.	NRC (1996)	*National Science Education Standards* 1996
Individuals are scientifically and technically literate: When their knowledge gives them a certain autonomy (the possibility of negotiating decisions without undue dependency with respect to others, while confronted with natural or social pressures; a certain capacity to communicate (finding ways of getting one's message across); and some practical ways of coping with specific situations and negotiating over outcomes.	Fourez (1997)	
The primary and explicit aim of the 5–16 science curriculum should be: To provide a course which can enhance 'scientific literacy' enabling students to express an opinion on important social and ethical issues with which they will increasingly be confronted.	Millar and Osborne (1998)	

continued

TABLE A-3 Continued

Definition	Source	Assessment/ Educational Standard
Since the 1950's there have been a variety of goals for teaching science and a wide range of meanings of scientific literacy: (1) teaching and learning about science as a cultural force in the modern world; (2) preparation for the world of work; (3) teaching and learning about science that has direct application to everyday living; (4) teaching students to be informed citizens; (5) learning about science as a particular way of examining the natural world; (6) understanding reports and discussions of science that appear in the popular media; (7) learning about science for its aesthetic appeal; (8) preparing citizens who are sympathetic to science; and (9) understanding the nature and importance of technology and the relationship between technology and science.	DeBoer (2000)	
The types of knowledge needed to engage in science in contemporary societies by individuals who are not professionally involved in science: (1) subject matter knowledge, (2) collecting and evaluating data, (3) interpreting data, (4) modeling in science, (5) uncertainty in science, and (6) science communication in the public domain.	Ryder (2001)	
Reading and writing when the content is science is the *fundamental* sense of scientific literacy, and being knowledgeable, learned, and educated in science is the *derived* sense. Scientific literacy comprises both the concepts, skills, understandings, and values generalizable to all reading, and knowledge of the substantive content of science.	Norris and Phillips (2003)	
A scientifically literate citizen needs to have (1) a basic vocabulary of scientific terms and constructs, and (2) a general understanding of the nature of scientific inquiry. The level of understanding needed for scientific literacy needs to be sufficient to read and comprehend the Tuesday science section of *The New York Times*.	Miller (2004)	

TABLE A-3 Continued

Definition	Source	Assessment/ Educational Standard
Scientific literacy refers to four interrelated features that involve an individual's: (1) scientific knowledge and use of that knowledge to identify questions, acquire new knowledge, explain scientific phenomena and draw evidence-based conclusions about science-related issues; (2) understanding of the characteristic features of science as a form of human knowledge and enquiry; (3) awareness of how science and technology shape our material, intellectual, and cultural environments; and (4) willingness to engage in science-related issues and with the ideas of science, as a constructive, concerned, and reflective citizen.	Bybee et al. (2009), OECD (2006)	PISA 2006
Indicators of science literacy are (1) a good understanding of basic scientific terms, concepts, and facts, (2) an ability to comprehend how science generates and assesses evidence, and (3) a capacity to distinguish science from pseudoscience.		NSB *Science & Engineering Indicators 2010*
Scientific literacy refers to an individual's: (1) scientific knowledge and use of that knowledge to identify questions, acquire new knowledge, explain scientific phenomena and draw evidence-based conclusions about science-related issues; (2) understanding of the characteristic features of science as a form of human knowledge and enquiry; (3) awareness of how science and technology shape our material, intellectual and cultural environments; and (4) willingness to engage in science-related issues, and with the ideas of science, as a reflective citizen.	OECD (2012a)	PISA 2012

continued

TABLE A-3 Continued

Definition	Source	Assessment/ Educational Standard
By the end of the 12th grade, students should have gained sufficient knowledge of the practices, crosscutting concepts, and core ideas of science and engineering to engage in public discussions on science-related issues, to be critical consumers of scientific information related to their everyday lives, and to continue to learn about science throughout their lives. They should come to appreciate that science and the current scientific understanding of the world are the result of many hundreds of years of creative human endeavor. It is especially important to note that the above goals are for all students, not just those who pursue careers in science, engineering, or technology or those who continue on to higher education.	NRC (2012)	*A Framework for K-12 Science Education*
The outcomes of scientific literacy can be categorized into three categories of values regarding (1) the states of *knowing* one might obtain, (2) the *capacities* one might refine, and (3) the *personal traits* one might develop.	Norris et al. (2014)	
Scientific literacy is the ability to engage with science-related issues, and with the ideas of science, as a reflective citizen. A scientifically literate person, therefore, is willing to engage in reasoned discourse about science and technology which requires the competencies to: 1. Explain phenomena scientifically: • Recognize, offer and evaluate explanations for a range of natural and technological phenomena 2. Evaluate and design scientific enquiry: • Describe and appraise scientific investigations and propose ways of addressing questions scientifically 3. Interpret data and evidence scientifically: • Analyse and evaluate data, claims and arguments in a variety of representations and draw appropriate scientific conclusions	Koeppen et al. (2008), OECD (2013) (draft)	PISA 2015
Key indicators of Americans' attitudes about and understanding of science and technology are (1) interest in new scientific discoveries, (2) basic scientific knowledge, (3) belief that science creates opportunity, (4) confidence in the scientific community, and (5) support for science funding.	NSB (2016)	*Science & Engineering Indicators* 2016

TABLE A-4 Health Literacy

Definition	Source	Assessment/ Educational Standard
Health literacy represents the cognitive and social skills which determine the motivation and ability of individuals to gain access to, understand, and use information in ways which promote and maintain good health. By improving people's access to health information, and their capacity to use it effectively, health literacy is critical to *empowerment*. Health literacy is itself dependent upon more general levels of literacy. Poor literacy can affect people's *health* directly by limiting their personal, social and cultural development, as well as hindering the development of health literacy.	WHO (1998)	
Health literacy is a constellation of skills, including the ability to perform basic reading and numerical tasks required to function in the healthcare environment. Patients with adequate health literacy can read, understand, and act on health care information.	AMA (1999), Davis et al. (1993), Parker et al. (1995), Baker et al. (1999)	REALM 1993, TOFHLA 1995, S-TOFHLA 1999
Assessment of various health-related activities from NALS 1992 and IALS 1994-1998 literacy surveys: 1) health promotion, 2) health protection, 3) disease prevention, 4) health care and maintenance, and 5) systems navigation.	Rudd et al. (2004)	HALS 2004
Health literacy is the degree to which individuals can obtain, process and understand the basic health information and services they need to make appropriate health decisions. But health literacy goes beyond the individual. It also depends upon the skills, preferences, and expectations of health information and care providers: our doctors; nurses; administrators; home health workers; the media; and many others.	IOM (2004), Weiss et al. (2005)	NAAL 2003, *Newest Vital Sign* 2005
The wide range of skills, and competencies that people develop to seek out, comprehend, evaluate and use health information and concepts to make informed choices, reduce health risks and increase quality of life.	Zarcadoolas et al. (2005)	

TABLE A-4 Continued

Definition	Source	Assessment/ Educational Standard
Health literacy goes beyond a narrow concept of health education and individual behavior-oriented communication, and addresses the environmental, political and social factors that determine health. Health education, in this more comprehensive understanding, aims to influence not only individual lifestyle decisions, but also raises awareness of the determinants of health, and encourages individual and collective actions which may lead to a modification of these determinants.	WHO (2009)	
A more comprehensive definition of health literacy must include both the abilities of individuals and the characteristics of professionals and institutions that support or that may inhibit individual or community action.	Rudd et al. (2012)	
Health literacy is linked to literacy and entails people's knowledge, motivation and competences to access, understand, appraise, and apply health information in order to make judgments and take decisions in everyday life concerning healthcare, disease prevention and health promotion to maintain or improve quality of life during the life course.	Sørensen et al. (2012, 2015), Pelikan et al. (2012)	European Health Literacy Questionnaire 2012
Components of a definition of health literacy should include (1) system demands and complexities as well as individual skills and abilities; (2) measurable components, processes, and outcomes; (3) potential for an analysis of change; and (4) demonstrate linkage between informed decisions and action.	Pleasant et al. (2016)	

TABLE A-5 Health Numeracy

Definition	Source	Assessment/ Educational Standard
Health numeracy is the degree to which individuals have the capacity to access, process, interpret, communicate, and act on numerical, quantitative, graphical, biostatistical, and probabilistic health information needed to make effective decisions.	Golbeck et al. (2005)	
Productive health information use results from the interplay between the quantitative competencies of the patient (health numeracy), the properties of the artifacts that mediate health cognition (information design), and the communication skills of the health-care provider.	Ancker and Kaufman (2007)	

Appendix B

Biographical Sketches of Committee Members and Staff

CATHERINE SNOW (*Chair*) is Patricia Albjerg Graham professor of education at the Harvard Graduate School of Education. Her work has been devoted to language and literacy development in children, focusing on how oral language skills are acquired and how they relate to literacy outcomes. Her research activities include a longitudinal study of language and literacy skills among low-income children who have been followed for 15 years since age 3; following the language development of young children participating in the Early Head Start intervention; studying the vocabulary development of first- and second-language learners; and considering aspects of transfer from first to second language in the domains of language and literacy. She has also been involved in work to develop consensus among teacher educators about what pre- and in-service elementary teachers need to know about language and literacy, as well as bilingualism and its relation to language policy issues. She is currently involved in efforts to improve middle-school literacy outcomes in partnership with the Boston Public Schools. She has a B.A. from Oberlin College in psychology and an M.A. and a Ph. D. in psychology from McGill University.

NICK ALLUM is professor of sociology at the University of Essex, where he directs the masters in science program in survey methods. His research encompasses survey methodology, public understanding of science, social and political trust, and risk perception. He currently serves as the general secretary for the European Survey Research Association, and previously served on the National Science Foundation's expert panel on science literacy indicators. He has previously also worked as a statistical consultant for the Pew Research Center, as well as performing survey design work for the United Kingdom's Department of Media Culture and Sport. He has a B.A. in political economy from the University of East London, an M.Sc. in social research methods from the London School of Economics, and a Ph.D. in social psychology at the London School of Economics.

EMILY BACKES (*Research Associate*) works mostly with the Committee on Law and Justice in the Division of Behavioral and Social Sciences and Education, providing substantive analysis, writing, and editing for studies on juvenile justice, forensic science, illicit markets, and policing. Previously, she worked with the Committee on Human Rights of the National Academy of Sciences, Engineering, and Medicine, where she was responsible for researching cases of unjustly imprisoned scientists worldwide and synthesizing scholarship on science and human rights issues. She has a B.A. and an M.A. in history from the University of Missouri, and she is pursuing a J.D. at the David A. Clarke School of Law of the University of the District of Columbia.

JOHN BESLEY is an associate professor and Ellis N. Brand chair in the Department of Advertising and Public Relations at the College of Communication Arts and Sciences at Michigan State University. He studies how views about decision makers and decision processes affect perceptions of science and technology with potential health or environmental effects, including consideration of both mediated exposure through newspapers, television programs, and web content and public engagement exercises (e.g., public meetings). His work explores the relationships between media use, public engagement, and health and environmental risk perceptions. His research has touched on public perceptions of nanotechnology, biotechnology, and energy technologies (particularly nuclear and hydrogen and fuel cell technologies). He has also been involved in research into journalistic norms related to coverage of public engagement and research to better understand the effects of science and risk communication training. He has a B.A. in journalism and an M.A. in public administration (innovation, science, and environment policy) from Carleton University in Canada and a Ph.D. in communications from Cornell University.

DOMINIQUE BROSSARD is professor and chair in the Department of Life Sciences Communication at the University of Wisconsin–Madison and is also affiliated with the university's Holtz Center of Science and Technology Studies and the Morgridge Institute for Research in the Center for Global Studies. She is a fellow of the American Association for the Advancement of Science and a former board member of the International Network of Public Communication of Science and Technology. Her work focuses on questions related to public understanding of science, with a specific emphasis on public opinion dynamics related to controversial scientific issues. She teaches courses in strategic communication theory and research, with a focus on science and risk communication. Previously, she held positions at Accenture in its Change Management Services Division and as the communication coordinator for the Agricultural Biotechnology Support Project II. She has an M.S. in plant biotechnology from the Ecole Nationale d'Agronomie de Toulouse and an M.P.S and a Ph.D. in communication from Cornell University.

KENNE DIBNER (*Study Director*) is a program officer with the Board on Science Education. Prior to this position, Kenne worked as a research associate at Policy Studies Associates, Inc., where she conducted evaluations of education policies and programs for government agencies, foundations, and school districts. Most recently, she concluded an evaluation of a partnership with the U.S. Department of Education, the National Park Service, and the Bureau of Indian Education to provide citizen science programming to tribal youth. Previously, she worked as a research consultant with the Center on Education Policy and served as a legal intern for the U.S. House of Representatives' Committee on Education and the Workforce. She has a B.A in English literature from Skidmore College and a Ph.D. in education policy from Michigan State University.

NOAH WEETH FEINSTEIN is associate professor in the Department of Curriculum and Instruction in the School of Education at the University of Wisconsin–Madison. His work explores the value of science in the social and political lives of nonscientists, including identifying and investigating the social mechanisms through which scientific institutions and practices can make societies more, rather than less, democratic. His current projects focus on public engagement with science among parents of recently diagnosed autistic children, the contribution of learning to climate change adaptation, the impact of changing scientific practices on scientists' outreach, and the need for museums and science centers to forge better connections with their diverse communities. He has a B.A. in biological sciences from Harvard University, an M.S. in biological sciences and neural development, and a Ph.D. in science education from Stanford University.

S. JAMES GATES, Jr., is a university system regents professor, the John S. Toll professor of physics, and director of the Center for String and Particle Theory, all at the University of Maryland. His work has long been in the fields of supersymmetry, supergravity, and string theory. He serves on the U.S. President's Council of Advisors on Science and Technology, and on the Maryland Board of Education. He is a member of the National Academy of Sciences, the American Academy of Arts and Sciences, and the American Philosophical Society. He is a fellow of the American Physics Society, American Association for the Advancement of Science, National Society of Black Physicists, and British Institute of Physics and a member of the board of trustees of the Society for Science & the

Public and the board of advisors for the U.S. Department of Energy's Fermi National Laboratory. He is a recipient of the Medal of Science, the highest recognition given by the U.S. government to scientists for his contribution to the mathematics of supersymmetry in particle, field, and string theories and his extraordinary efforts to engage the public on fundamental physics. He has B.S. degrees in mathematics and in physics and a Ph.D. in physics, all from the Massachusetts Institute of Technology.

LOUIS GOMEZ holds the MacArthur chair in digital media and learning in the Graduate School of Education and Information Studies at the University of California, Los Angeles. He is also a senior partner at the Carnegie Foundation for the Advancement of Teaching, where he leads the Network Development work. Previously, he held positions at Northwestern University and at the University of Pittsburgh, where he directed the Center for Urban Education and was a senior scientist at the Learning Research and Development Center. Prior to joining academia, he worked in cognitive science and person–computer systems and interactions at Bell Laboratories, Bell Communications Research, Inc., and Bellcore. His research interests have encompassed the application of computing and networking technology to teaching and learning, applied cognitive science, human–computer interactions, and other areas. He is a member of the National Academy of Education. He has a B.A. in psychology from the State University of New York at Stony Brook and a Psy.D. in cognitive psychology from the University of California, Berkeley.

ALEXA MCCRAY is professor of medicine at Harvard Medical School. She conducts research on knowledge representation and discovery, with a special focus on the significant "Tower of Babel" problems that persist in the curation, dissemination, and exchange of scientific and clinical information in biomedicine and health. Previously, she was director of the Lister Hill National Center for Biomedical Communications, a research division of the National Library of Medicine at the National Institutes of Health; on the research staff of IBM's T.J. Watson Research Center; and on the faculty of Harvard Medical School. She is a member of the National Academy of Medicine. She is a fellow of the American Association for the Advancement of Science and a fellow of the American College of Medical Informatics (ACMI). She is the immediate past president of ACMI and is a past member of the board of both the American Medical Informatics Association and the International Medical Informatics Association. She has a B.A. in modern languages from Skidmore College, an M.A. in German literature and language from Boston College, an M.S. in linguistics from Georgetown University, and a Ph.D. in linguistics from Georgetown University.

JANET OHENE-FREMPONG is president of J O Frempong & Associates, Inc., which provides a range of communication services, including consumer research, materials and forms development, program development, presentations, seminars and institution-based coaching in consumer health communications. Formerly, she was director of the Health Literacy Project at the Health Promotion Council of Southeastern Pennsylvania, and she has conducted workshops and provided consultation on plain language and cross-cultural communication for a wide range of health information providers, including health care systems, government agencies, health insurers, pharmaceutical companies, biotechnology companies, medical publishers, health and human service agencies as well as schools of medicine, nursing and allied health. She has served on a number of national boards and advisories, is a founding and emerita member of the Clear Language Group and is an Institute for Healthcare Advancement Strategic Partner. She has a B.A. in political science from Cornell University and an M.S. in public health nutrition from Columbia University Teachers College.

JONATHAN OSBORNE holds the Kamalachari chair in science education at the Graduate School of Education at Stanford University. Previously, he held the chair in science education at King's College London, and he served as an advisor to the U.K. House of Commons Science and Technology Committee for its report on science education. Currently he chairs the expert group that produced the framework for the science assessments conducted by the OECD Programme for International Student Assessment. His research interests are in the role of argumentation in science and improving the teaching of literacy in science. He has a B.Sc. in physics from Bristol University, a postgraduate certificate in education from Cambridge University, a master's degree in astrophysics from Queen Mary College at the University of London, and a Ph.D. in education from King's College at the University of London.

JULIE ANNE SCHUCK (*Associate Program Officer*) has provided analytical, administrative, writing, and editorial support for a wide range of studies in the Division of Behavioral and Social Sciences and Education, mostly for the Committee on Law and Justice. In addition to her work on this study, her recent projects have included studies of the science of human-system integration; science, technology, engineering, and mathematics education; incarceration in the United States; reforming juvenile justice; understanding the U.S. illicit tobacco market; strengthening the National Institute of Justice; and support for forensic science research. She has an M.S. in education from Cornell University and a B.S. in engineering physics from the University of California, San Diego.

HEIDI SCHWEINGRUBER (*Senior Research Associate*) is the director of the Board on Science Education at the National Academies of Science, Engineering, and Medicine. She has served as study director or co-study director for a wide range of studies, including those on revising national standards for K-12 science education, learning and teaching science in grades K-8, and mathematics learning in early childhood. She also coauthored two award-winning books for practitioners that translate findings of Academies' reports for a broader audience, on using research in K-8 science classrooms and on information science education. Prior to joining the Academies, she worked as a senior research associate at the Institute of Education Sciences in the U.S. Department of Education. She also previously served on the faculty of Rice University and as the director of research for the Rice University School Mathematics Project, an outreach program in K-12 mathematics education. She has a Ph.D. in psychology (developmental) and anthropology and a certificate in culture and cognition, both from the University of Michigan.

EUGENIE C. SCOTT was the founding executive director of the National Center for Science Education (NCSE) and now serves on NCSE's advisory council. She has long been both a researcher and an activist in the controversy about creationism and evolution and addressed many aspects—educational, legal, scientific, religious, and social—of the controversy. She has received national recognition for her NCSE activities, including awards from scientific societies, educational societies, skeptics groups, and humanist groups. She holds nine honorary degrees and is the recipient of the National Academy of Sciences' Public Welfare Medal. She has a B.S. and an M.S. in physical anthropology from the University of Wisconsin–Milwaukee and a Ph.D. in physical anthropology from the University of Missouri.

EARNESTINE WILLIS is a Kellner professor in pediatrics, director of the Center for the Advancement of Underserved Children in the Department of Pediatrics, and holds an appointment in the Institute of Health & Society, all at the Medical College of Wisconsin. Previously, she served as director of a Federally Qualified Look-A-Like Health Center at the University of Chicago and as chair of the State of Wisconsin Tobacco Control and Prevention Board. Her work has focused on documenting and advocating the social and health needs of children in underserved neighborhoods, and she has also led numerous successful initiatives for school-based health services; avoiding adverse birth outcomes through a community-based life course perspective; early literacy promotion; oral health improvement; child advocacy; emergency preparedness; and supporting lactation in the workplace for women returning to work. She received an honorary Ph.D. from Cardinal Stritch University and the Community Impact Award from the Medical Society of Milwaukee County. She has a B.S. from Tougaloo College, an M.D. from Harvard Medical School, and an M.P.H. from Harvard School of Public Health.